Praise for

Raising GOATS
Naturally

With endearing personal stories and layman's scientific explanations, *Raising Goats Naturally* lays an enjoyable and empowering foundation for goat-rearing success on the self-reliant farmstead. Deborah Niemann exemplifies the best spirit and action in homestead animal care. What a great contribution to self-reliance.

— Joel Salatin, Polyface Farm

Deborah Niemann's book is an inspiring and useful guide for anyone thinking about raising goats. Her research is exhaustive and her personal stories give real depth and dimension to the experience, preparing the prospective goat owner not only for the technical challenges, but also for the rich emotional experience of sharing your life with a herd.

—Bryan Welch, Publisher & Editorial Director, *Mother Earth News, GRIT Magazine, Mother Earth Living*, and author, *Beautiful and Abundant: Building the World We Want*

Deborah Neimann has done it again! She has created another practical and educational book for beginners interested in trying something new. *Raising Goats Naturally* is an excellent starting point and guide for anyone about to dive into the world of goats. The book does a marvelous job in covering practical approaches to husbandry and how to avoid the pitfalls that novices often encounter.

—Jeannette Beranger, Research & Technical Programs Manager, American Livestock Breeds Conservancy

In *Raising Goats Naturally*, Deborah Neimann cuts through the formulaic and often inflexible so-called "expert advice" and encourages us to get to know our animals and listen to what they tell us. Drawing on vast experience, Neimann offers an upbeat, authentic glimpse of what life with dairy goats is really like. This book is important because it brings animal husbandry back to the fore and delivers the goods in a highly integrated manner that's every bit as enjoyable to read as it is important for goatherds of all experience levels.

—Oscar H. "Hank" Will III, Editor-in-Chief, *GRIT Magazine*
and author, *Plowing with Pigs*

Deborah Niemann's conversational style and storytelling make this more than a book about goats, it's a look into a life shared with these amazing animals and how it is bettered by their presence. A wonderful guide for novices and the seasoned alike!

—Jenna Woginrich, author, homesteader and blogger

Raising GOATS
Naturally

Today, more than ever before,
our society is seeking ways to live more conscientiously.
To help bring you the very best inspiration and information
about greener, more sustainable lifestyles, *Mother Earth News*
is recommending select books from New Society Publishers.
For more than 30 years, *Mother Earth News* has been
North America's "Original Guide to Living Wisely,"
creating books and magazines for people with a passion
for self-reliance and a desire to live in harmony with nature.
Across the countryside and in our cities, New Society Publishers
and *Mother Earth News* are leading the way to a wiser,
more sustainable world. For more information,
please visit MotherEarthNews.com.

Raising GOATS
Naturally

THE COMPLETE GUIDE TO
MILK, MEAT
AND MORE

Deborah Niemann

new society
PUBLISHERS

Cover design by Diane McIntosh.
Cover images: Goats courtesy of Katherine Boehle; Cheese—Deborah Niemann
Interior goat illustrations © Melissa Kruger, www.melissakruger.com.

Printed in Canada. First printing September 2013

Paperback ISBN: 978-0-86571-745-9   eISBN: 978-1-55092-543-2

Disclaimer

The information contained in this book is not intended to replace veterinary advice. It is based on the experience of the author and named contributors in raising goats and as such is anecdotal. All information about drugs, use, extra-label use, dosage, method of administration, and any other specific drug use, is anecdotal and is presented for discussion only. It should not be construed as veterinary advice, nor does it replace advice of a licensed veterinarian.

Inquiries regarding requests to reprint all or part of *Raising Goats Naturally* should be addressed to New Society Publishers at the address below.

To order directly from the publishers, please call toll-free (North America) 1-800-567-6772, or order online at www.newsociety.com

Any other inquiries can be directed by mail to:

New Society Publishers
P.O. Box 189, Gabriola Island, BC V0R 1X0, Canada
(250) 247-9737

LIBRARY AND ARCHIVES CANADA CATALOGUING IN PUBLICATION

Niemann, Deborah, author
Raising goats naturally : the complete guide to milk,
meat and more / Deborah Niemann.

Includes bibliographical references and indexes.
ISBN 978-0-86571-745-9 (pbk.)

1. Goats. I. Title.

SF383.N53 2013   636.3'9   C2013-904397-7

New Society Publishers' mission is to publish books that contribute in fundamental ways to building an ecologically sustainable and just society, and to do so with the least possible impact on the environment, in a manner that models this vision. We are committed to doing this not just through education, but through action. The interior pages of our bound books are printed on Forest Stewardship Council®-registered acid-free paper that is **100% post-consumer recycled** (100% old growth forest-free), processed chlorine free, and printed with vegetable-based, low-VOC inks, with covers produced using FSC®-registered stock. New Society also works to reduce its carbon footprint, and purchases carbon offsets based on an annual audit to ensure a carbon neutral footprint. For further information, or to browse our full list of books and purchase securely, visit our website at: **www.newsociety.com**

Contents

Acknowledgments

It was a happy day when Ingrid Witvoet of New Society Publishers asked me if I'd ever thought about writing a book about goats. It is always a blessing when one can combine two great passions in a single project, and there are few things in this world that I love more than writing and goats. I will forever be grateful to the wonderful people at New Society for making this book possible. And as always, many thanks to my editor Janice Logan for her invaluable insight and expertise.

One of the best things about writing this book was the opportunity to tap the brains of experts, who know far more than I do on various aspects of goat care. I've been reading the research of goat extension specialist Steve Hart, PhD, of Langston University for years, and it was an honor to have him provide feedback on the parasite section of this book. I also want to offer a big thank-you to Sandra G. Solaiman, PhD, PAS, Professor Emeritus and Adjunct Professor at Tuskegee University and Affiliated Professor at Auburn University for sharing her knowledge of goat nutrition.

I could not have written this book were it not for the members of the NigerianDwarfGoats.ning.com community who have shared their challenges, frustrations, knowledge, and successes over the years. Reading their stories made me realize the variety of problems that goat owners face on a daily basis, as well as the diversity of solutions. I continue learning from them every day. I also owe a special debt of gratitude to the goat owners who were willing to share their stories and passion specifically for this book.

And of course, I never would have had the time to write this book if it were not for the countless contributions of my family, especially my husband Mike, who took over many of the milking and cheese-making duties, and my son Jonathan, the only child still at home, who took on almost all of the family cooking duties, as well as more farm chores. Thanks also to Katherine for pitching in during college breaks and to Margaret for her continuing moral support. Both of my daughters also deserve a special thank-you for growing our herd beyond the "couple of does for making goat cheese" that I had originally envisioned. Because of their desire to show and be on milk test, we wound up with much better genetics than if they had never become involved. I know their presence as co-owners of the goats reduced the learning curve for me considerably.

Introduction

※ ※ ※

IT WAS LOVE AT FIRST BITE when I tasted goat cheese at a party in Vermont when I was nineteen years old. More than a decade passed before I saw goat cheese in a grocery store and immediately snatched it up. But at a dollar an ounce, it was a rare treat. When my husband and I started talking about moving to the country to grow our own food organically, goats were one of the three species of livestock I wanted, and I wanted them simply for that delicious cheese. I wanted chickens for eggs and cows for milk and butter.

While the chickens and goats proved to be easy for a city slicker to learn to raise, the cows were a different story. I had purchased Irish Dexters, which are the smallest breed of cattle, but I quickly learned that it really didn't matter whether a cow weighed 800 pounds or 1500 pounds—if she wanted to do something, she could easily get her way because she outweighed me by so much that it was hopeless.

I made similar mistakes in purchasing both the cows and goats. I bought animals that had no experience with milking, and I didn't even buy animals whose mothers had been milked, so the genetic potential as milkers was a mystery. I made the novice mistake of assuming that because every female mammal makes milk, they would all make good milkers. I knew nothing about udder texture, teat size, orifice size—

1

I didn't even know what an orifice was! And it never occurred to me that a cow or goat might not be terribly excited at the prospect of being milked.

Learning to milk the goats was not the easiest thing I ever did, but it wasn't terribly difficult either. It was Mother's Day 2002 when I brought home my first two goats, a two-month-old doeling and an unrelated three-year-old doe that had been nursing triplets. My husband had built a milk stand based on pictures we found on the Internet. I arrived home with the goats shortly before sundown and attempted to milk Star, the three-year-old. We put her on the milk stand and filled up the feed bowl. She took a couple of bites, but as soon as I touched her udder, she kicked the bucket, turned her head around, and glared at me. She continued to give me this look that I translated as, "*WHAT* are you do-ing back there?" There were ultimately four of us working towards the single goal of extracting milk from this goat. My husband held her back legs so she couldn't kick over the bucket. My two daughters scooped up the grain in their hands and sweet talked her, saying, "Here, Star, don't you want some yummy grain?" The goat continued to glare at me. Then I remembered reading somewhere that music relaxes animals and that some people play music in their milking parlors, so I suggested that we sing. "Twinkle, Twinkle Little Star" seemed appropriate given the goat's name, but she was not impressed. However, within a few days I was able to milk Star by myself with no one holding her legs or sweet talking her or even singing. It was my first lesson in the importance of the three Ps: practice, persistence, and patience.

The cows were a completely different story, though. I was never able to even touch their udders. Despite the fact the seller had said they would be very easy to train—"Just tie 'em up for a couple of days, and they'll be following you around like a dog"—they were range cows, never handled during their first year of life before I purchased them. Although livestock are domesticated, they have to be handled from the time they're born, or they can easily return to a feral state of mind. We wound up selling our first two cows after a couple of years, but I also came to the realiza-tion that we didn't need cows. The goats could meet all of our dairy needs—and more.

That soft creamy cheese that so many people call "goat cheese" is more correctly called chèvre (pronounced like "shev"), and it is possible to make many types of cheese and other dairy products from goat milk. The first cheese I made was queso blanco, and it was quickly followed by chèvre, yogurt, kefir, and queso fresco. A few months after starting to make cheese, I began to make goat milk soap. Eventually we started making aged cheeses, and for the past few years, we have made 100 percent of the cheese that our family uses, including cheddar, mozzarella, Parmesan, Gouda, Havarti, and more. Although we were vegetarians when we started our homesteading adventure, today we also eat goat meat and use goat leather. Even our goats' manure contributes to our homestead, as it is the only fertilizer we use in our garden.

It makes a lot of sense to raise goats for milk production for your family because goats are smaller than cows, eat less, poop less, are easier to handle, and produce a more manageable amount of milk. A potential buyer called me a few years ago because after a couple of years with a cow, her family realized that they didn't need the amount of milk a cow produced. They were not interested in making cheese, so it made no sense for them to have an animal that was producing 5 gallons of milk a day. Because dairy animals are all herd animals, you always need to have at least two animals, and with goats it is easy to add to your herd, especially when that special kid is born that you just can't bring yourself to sell. "Just one more goat" doesn't eat nearly as much as "just one more cow."

My journey with goats has been an interesting one, generally made easier and only sometimes more complicated by the Internet. I joined online groups and forums where people would answer my questions when I came across a situation that was not answered in any of my books. In the early 2000s, most of the people answering questions on the groups had been breeding goats for at least a few years and had a lot of good information. Today, however, because goats have become more common, there are a multitude of websites and blogs putting out information, some of which is questionable or downright wrong. Although information is more plentiful than ever, it is also more challenging to weed through it all to get accurate information.

There is no one-size-fits-all approach to raising goats. When I was in graduate school working towards my master's degree in communication, I had a professor who would often throw out a question and after someone gave an answer he'd nod and then ask, "Anyone else?" Someone would hesitantly raise a hand and say, "Well, it depends." The professor would smile and respond, "That's the grad school answer." He would reassure the first person that their answer was not necessarily wrong and point out that there could be multiple right answers to the question, depending on the situation. This is often the case when raising goats. Many people want to know exactly what to feed, what supplements to use, and whether a management practice is safe. Usually the answer is, "It depends."

The goal of this book is not to put forth the single best plan for raising goats and making dairy products. Quite simply, the best plan on my farm probably won't work for most other farms. It should be obvious that goats on the Illinois prairie will require different management

---

❖ JULIANA GOODWIN, Punta Gorda, Florida

We started our little endeavor partly because I'm horrified by factory farm treatment of animals and partly because I think a lot of the food that is being mass marketed right now is very unhealthy to downright poisonous.

When I started raising our chickens and goats I had an epiphany about the "cost" of food. And I don't really mean money. Huge chain stores advertise "cheap" food, but I think the idea of "cheap" meat, eggs, or milk is an insult. There is nothing cheap about life. The amount of waste in this country generated either by individuals overindulging or restaurants or other institutions throwing food away affects the real cost of food, just as do poor management practices in the mass production of meat or eggs that cause huge recalls and the disposal of thousands of pounds of these products.

I have learned by watching pregnant does waddle around, scream in labor and go through everything they go through to make milk that it's a big deal. It isn't just some beverage that appears in a bottle at the grocery store: an animal carried a baby, delivered and loved that baby, and then put their life energy into making that milk. I can't stand to waste an egg or a cup of milk that I and my animals have labored over producing (pun intended). An enormous amount of collective effort, animal and human, has gone into that egg or milk and it is special.

from those in the Arizona desert or the mountains of British Columbia. But if the farmer two miles from me raised goats, they would require different management as well because the well water on that farm does not have the high sulfur content of my well. If I'm starting to lose you because it sounds like goats might be too complicated, wait! It really isn't complicated.

The goal of this book is to give you the information that you need so that you can figure out what will work best for you and your goats. I see a lot of new goat owners online asking why they see so much contradictory information and wanting to know who is right. Is a certain brand of mineral the best? Should you give injectable mineral supplements? Why can one person's goats do well with a mineral block while other goats need loose minerals? The reality is that sometimes two people with seemingly opposite ideas are both making the right decision for their goats. This is why it is important for you to understand the "why" behind the recommendations. If you simply try to duplicate the practices

To associate "cheap" or "disposable" with this milk is to say that my little goat's life, love for her baby, and effort to make milk is not worthy of the dignity we generally assign to living beings. I think that separating the food product from the intimate relationship with the living being that produced it is what allows us to treat factory-farmed animals so terribly.

Around the world I see that some animals are afforded a certain quality of life or protection under the law, such as pet dogs, and some are not, such as factory-farmed animals. Some people are afforded certain rights and some people are deprived of these rights due to societal prejudice. It is my personal philosophy that no

life, human or animal, is cheap or expendable. My greatest hope for our farm is that my human children will grow up with an enduring respect for all life. I hope they know there is not a type of animal, breed of animal, or use for an animal that justifies forcing that animal to live with zero dignity or respect. I believe that this sort of respect for animal life will also help them to understand that there are no "types" of people who are less deserving of any quality of life.

of some award-winning herd, you could wind up with dead goats, and that is not an exaggeration or a hypothetical conclusion. It has happened.

Throughout this book there are stories that tell you about what various goats have taught me. I've done this because I truly believe that I have learned far more from my goats than from any book, website, or veterinarian. Your goats will let you know whether your management style is working for them. This book will give you a good basic knowledge of goats' needs, but ultimately it is by listening to your own animals that you will figure out the best way to care for them. When a goat gets sick, has difficulty birthing, or dies, it has just given you valuable information about your management practices and possibly about its own genetics. It is also giving you information when the fertility rate skyrockets and milk production goes up. Whether a kid grows quickly or slowly, it is giving you information about its mother's milk production. This book will help you understand what the goats are telling you so that you can provide them with the environment and diet that will help them reach their genetic potential.

You may be wondering what "raising goats naturally" means. It is definitely not what happens in factory farms, but it is not strictly organic either. It is important to understand that under organic standards an animal cannot be denied medical attention. The animal is supposed to be treated with conventional medication when necessary, but its milk cannot be sold as organic for the rest of the current lactation. When a meat animal is treated with conventional medication, it can never be sold as organic. There is no legal definition of "natural" food, but in my world it means that animals are not given antibiotics in their daily rations and they are not injected with hormones to increase milk production or to get bred. They are not given dewormers on a regular basis—either chemical or herbal. Just as it is my personal goal to have a diet and lifestyle that allow me to stay healthy and avoid routine medications, my goal for my herd is that they will stay healthy with the proper diet and management.

Goats have enriched my life in so many ways—from their charming personalities to their delicious cheese. Unfortunately, goats have a bad reputation—undeserved, in my opinion—for being difficult to handle and having off-flavored milk. And some people wrongly assume that having a dairy animal is sentencing you to twice-daily milkings every day of the year with no holidays. So another goal of this book is to dispel misconceptions about goats.

Whether you are just thinking about getting a couple of goats to make your own cheese or you are further along in your personal goat journey, there is always more to learn. Every goat is an individual and will present you with its own unique personality and physical traits. The milk that you get from month to month will be a little bit different,

---

The Question of Lactose

Can I drink goat milk if I'm lactose intolerant? The answer to this question is actually quite complicated. Many people assume any type of physical discomfort following milk consumption is due to lactose intolerance, but there are a number of reasons why you may have difficulty drinking milk. If you are truly lactose intolerant, you cannot comfortably consume any milk because all milk contains lactose, a milk sugar. Aged cheeses will have less lactose in them as they age, so the older the cheese gets, the less you may react to it.

A true milk allergy, however, is a reaction to the milk protein, and this can vary when consuming the milk from one species to another, so you might react negatively to cow milk but be able to drink goat or sheep milk. There are people who have difficulty digesting pasteurized milk but are fine with raw milk. And then there is the most confusing group— those who react negatively to dairy products only sometimes. They may be reacting to the drugs or hormones that are in the milk, which can vary from day to day, depending upon whether the milk came from a farm that uses hormones or when a cow received her last dose of an antibiotic. Even though no detectable level of antibiotics is permitted in milk for sale, a sensitive person may react to residual levels of antibiotic that are below what is detectable by modern testing procedures.

If you cannot happily consume milk and dairy products, try goat milk or goat cheese before actually buying a couple of goats to make sure that you will be able to eat and drink your homegrown products.

providing you with cheese-making surprises. Like every other aspect of living a self-reliant lifestyle, you can't expect perfection. But at some point you realize that perfection really is not the goal.

The reason you have goats on your homestead is not necessarily to produce the perfect cheese or to create a million dollar corporation making artisanal goat cheese. Goats on your homestead provide you with milk that is fresher than anything money can buy. It comes from animals that spend their days outside in the sunshine breathing fresh air. It comes from animals that have names and are loved and cared for. They are not given hormones to increase milk production or to grow faster than nature intended. Your homegrown meat and your homemade dairy products are free from ingredients that you can't pronounce. Although homestead goats can save you money, the reality is that the benefits are priceless.

PART I

Planning, Purchasing, and Protecting

If you grew up consuming cow milk, you may have considered a cow when you decided to start producing your own dairy products. But there are plenty of reasons why goats are a better option for most people in modern society. Goats are easier to handle simply because they are smaller than cows. If you did not grow up on a farm, where you got used to handling cattle, goats will be less intimidating. It can be almost impossible to find a trained milk cow to purchase, but training a goat is not as difficult or as potentially dangerous for the novice as training a cow that has never been milked. It is also less expensive to get started with goats because they do not require the heavy-duty handling equipment needed for safe handling of cattle.

Although goats are easier to raise than cattle, this does not mean that you can just bring them home and let them run free in the pasture and expect all of their needs to be met. This section will give you the information you need to consider before getting goats as well as information on choosing a breed, on housing, bedding, fencing, livestock guardians, and more so that you have everything in place and ready when you bring your goats home. If you already have goats, this section might give you ideas for making your life easier or your goats happier.

CHAPTER 1

Choosing Your Goats

✢ ✢ ✢

AFTER DECIDING that I wanted Nigerian Dwarf goats, mostly because they were listed on the American Livestock Breeds Conservancy conservation list, I bought the first three that I found for sale. As you might imagine, there is a better way to go about choosing goats. It never occurred to me that some might be better milkers than others in terms of production, personality, or mammary system. Those are just a few of the things to take into consideration before buying.

How much milk do you want every day for consumption as fluid milk? How much cheese do you want to make? Do you want to butcher extra bucks for meat? How much meat do you want? Do you want fiber? Can you handle a 200-pound animal, or do you need one around 75 pounds? By the time you finish reading this section, you should have figured out how many goats you need and narrowed down the breed options, and you will have a good idea how to find goats that will meet your needs.

Breeds

There is a lot to consider when choosing a breed of goat, and it goes far beyond the descriptions of their color, personality, and milk production. The following information about the different breeds can serve as a starting point.

This mini-LaMancha has the characteristic elf ears of a full-sized LaMancha, but it is several inches shorter. Production and butterfat fall somewhere between that of a Nigerian Dwarf and a LaMancha. Many people are drawn to mini-LaManchas and mini-Nubians because they like the non-erect ears but prefer the smaller size.

There are eight breeds of dairy goats common to the United States and Canada: Alpine, LaMancha, Nigerian Dwarf, Nubian, Oberhasli, Saanen, Sable, and Toggenburg. The Guernsey is a rare breed that is slowly increasing in number in North America. All of these are standard-sized except for the Nigerian Dwarf.

There are also miniature dairy goats, which are hybrids of the Nigerian Dwarf and any of the standard-sized breeds. In order to avoid birthing difficulties when breeding for a hybrid, the buck must be the Nigerian Dwarf and the doe must be the standard-sized goat. The hybrids are referred to as the mini-Alpine, mini-Nubian, and so on. Although Pygmies used to be raised for dairy, the focus of most breeders has turned towards raising them for pets in the last couple decades, meaning that milk production and ease of milking are not emphasized.

You may also see "grade" or "experimental" goats, which are usually crossbreeds. A "recorded grade" is a goat whose pedigree is recorded with the American Dairy Goat Association (ADGA) but is not registered as a purebred.

The production and butterfat averages listed in the breed descriptions are from the American Dairy Goat Association, which keeps milking records for herds that are on Dairy Herd Improvement (DHI), which means the goats are milked once a month under the supervision of a milk tester. The milk is weighed, and a sample is sent to a lab where it is tested for butterfat, protein, and somatic cell count. ADGA keeps track of the milking records so that breeders can see how their goats measure up to others in the breed. Some might argue that goats on test will have higher average production than goats not on test because only breeders with exceptional producers will want to test.

Alpine

Sometimes called the French Alpine, this breed comes in a variety of colors and patterns. They have erect ears and a straight nose. The does should be at least 30 inches tall at the withers and weigh at least 135 pounds. Bucks should be at least 32 inches tall and weigh at least 170 pounds. The Alpine's butterfat averages 3.3 percent and production is around 2400 pounds of milk over a nine to ten month lactation. Alpines are a popular breed for those who want a lot of fluid milk, including commercial goat dairies.

Valium is an Alpine doe at Triple Creek Dairy in Iowa. Her color is "cou blanc," which means white neck. The Alpine's striking colors and markings are one reason for the breed's popularity.

Guernsey

The Guernsey is a recent addition to the dairy goat scene in the United States. The breed is being developed from Golden Guernsey embryos that were imported in the 1990s. Those offspring as well as some imported semen were crossed with Swiss-type dairy goats here. The Guernsey is medium-sized, similar to the Oberhasli or Toggenburg. Guernseys are critically endangered worldwide, which attracted the interest of Teresa Casselman of Six Point Farm in Bloomington, Illinois, who has been raising Nubians since 1994.

"I first learned about the Guernsey breed in 2003 when the *Dairy Goat Journal* featured the Golden Guernsey goat on its cover. As the name implies, the Golden Guernsey goat originated on the Island of Guernsey and nearby Channel Islands," said Teresa. "I continued to follow the progress of the breed in the United States, and in 2011 I purchased my first Guernsey does. By this time, both does and bucks were starting to become available, but they were still few and far between. I drove to Pennsylvania for my does and to Washington for my buck. The does

Snowbird Angelo is a Guernsey buck. Although both does and bucks grow beards, you may see pictures of does without beards because they are cut off when does are clipped for shows. (Photo courtesy of Teresa Casselman.)

were bred and kidded in 2012. As beginner's luck would have it, my first Guernsey kidded with quad does."

Teresa describes Guernsey goats as having a friendly and affectionate temperament. "Many people," she said, "are attracted to their golden hair coats, which can be short or long and flowing and range in color from pale cream to deep russet." She believes that the Guernsey breed's "productivity and smaller size make them ideal for a household or a less intensive production system."

Because Guernseys are still new to this continent, official milk production and butterfat averages are not yet available.

LaMancha

The LaMancha is the only dairy goat that claims the United States as its home. Its history dates back only about a century, unlike many of the European breeds, which have been around for hundreds of years. The distinguishing characteristic of the LaMancha is the ears—or lack thereof. I had LaManchas for seven years, and typically the first thing

This LaMancha doe and her buckling show the diversity of color available in the breed.

anyone asks when they see one for the first time is, "What happened to its ears?" Gopher ears are supposed to be almost non-existent up to 1 inch in length, whereas elf ears can be up to 2 inches long. Although does can have either type of ears, bucks can only be registered if they have gopher ears. LaManchas may be slightly smaller than Alpines, but not more than a couple of inches. LaManchas average 2200 pounds of milk with 3.8 percent butterfat.

Jackie Kuni of Alcmene LaManchas in Clarksville, Tennessee, fell in love with the ears, but she also likes the breed's loving personality and will to milk as well as the wide variety of colors available. "I really have no idea why I fell in love with them above all others but I did, and hard!" Jackie says.

Nigerian Dwarf

Many Nigerian Dwarf owners originally chose the goat for its small size or its high butterfat, or perhaps both. The maximum height is 22.5 inches for a doe and 23.5 inches for a buck in order to be registered with the American Dairy Goat Association or American Goat Society (AGS). Sometimes confused with Pygmy goats because of their small size, the Nigerian Dwarf is a small dairy goat and has a very different body type from the Pygmy, which has more of a meat goat body type and does

In spite of their small size, Nigerian Dwarf goats do well in cooler climates.

not produce as much milk as a Nigerian Dwarf. I once heard a judge say that the ideal Nigerian should look like someone took a picture of an Alpine or a Saanen and shrank it on a copy machine. The average Nigerian Dwarf produces 715 pounds with 6.5 percent butterfat, making it an excellent choice for those who want to make cheese.

Nubian

The Nubian, whose history goes back to Asia, Africa, and Europe, has two distinguishing characteristics that set it apart visually from the other standard-sized dairy goats—its long, pendulous ears and its Roman nose.

The Nubian is also unique in its butterfat, which tends to be higher than in the other standard-sized breeds, although milk production tends to be lower, averaging 1750 pounds at 4.7 percent butterfat. "The higher butterfat and protein are great for my cheese making," says Brendia Kempf, who has Nubians in her herd at Triple Creek Dairy in Iowa.

Keep in mind that every goat owner has their own reasons for preferring a particular breed. What appeals to one breeder might not appeal to you. "What I do like about Nubians is what most people would put on their 'don't like' list," says Vicki McGaugh of Cleveland, Texas. "They are bossy, they are loud, they have distinct personalities, and living just ten acres away from thousands of acres of national forest, they are fearless."

Tasmania is pictured here as a doeling at Triple Creek Dairy. The Nubian's Roman nose, which is convex rather than straight or dished, is a disqualification in any other breed of dairy goat. When the ears are held flat against the face, they should extend at least 1 inch beyond the end of the muzzle.

While some people may be drawn to less common breeds, Vicki likes the Nubian's popularity. "The bloodlines are so diverse, the herd book so large, that you can really breed this breed into whatever you like it to be."

Oberhasli

The Oberhasli was originally called the Swiss Alpine in the United States, and the breed was registered in a sub-herd book of the Alpine breed until the late 1970s.

"One of the traits that originally attracted me to the breed in 1991 was that they are a more moderate size than most of the other Swiss or erect eared breeds," says Tom Rucker of Buttin'Heads Dairy Goats in Marengo, Ohio. Although there is no upper limit on size, the minimum size is 2 inches shorter than the Alpine. The Oberhasli buck must be chamoisee, which is red with black markings, although does can be chamoisee or black.

❖ ELLEN F. DORSEY, Dill's-A Little Goat Farm and Dorsey-Lane Nubians, Chelsea, Oklahoma

Over the years, I've owned several breeds of dairy goats, including the controversial Pygmy goat (is it a dairy goat or isn't it?!). I finally settled on three breeds. Currently I raise Nigerian Dwarves as my main breed, Alpines and Nubians. Why? I suppose there are a variety of reasons.

The Nubian—I love to watch them regally cross the pasture. None of the breeds quite measure up to the Nubian in its gait or stance. They have nice butterfat and protein numbers and produce very sweet milk. Mine, however, are quite dumb and try my patience on a daily basis. You see, I am a rather energetic person, so I am constantly cleaning or fiddling with things at milking time, trimming hooves or treating some ailment or another. A simple relocation of the

broom will result in an entire group of Nubians refusing to enter the milk parlor, which means I must drag them in one at a time, increasing my already long chore schedule.

The Alpine—I'm not quite sure why I have Alpines. I guess I just like them! I do love the look of the Swiss breeds, and this one allows pretty much any color or color combination, which certainly is eye appealing when scanning the pasture. My Alpines are high producing does with a generic tasting milk. Not sharp and goaty, but not the truly sweet milk that my Nigerian Dwarves produce. I've found, because I sell milk to whitetail deer farmers, that a combination of the high butterfat/protein milk of the Nigerian cut with Alpine milk seems to be perfect for good

SGCH Buttin'Heads Sofia *M is an Oberhasli. The letters around her name mean that she is a finished champion in ADGA with a milk star and a superior genetics designation. (Photo courtesy of Tom Rucker.)

growth patterns in deer fawns, and you cannot reproduce this combination using a strictly Nubian herd. The Alpine typically has a fun personality. One minute she's pawing and snorting at an enemy; the next, she's climbing the walls to get away! She has a fight or flight instinct that in my herd is confused at best!

The Nigerian Dwarf—aside from their easy-to-manage smaller size (there isn't a Nigerian on the place that I cannot pick up and move to where I want it when necessary!) they have a personality that is bigger than life. They rule on this farm. Even the staunchest, most heroic of Alpines will turn tail and run when met with a chorus of angry Nigerian Dwarves with hackles raised! They have incredibly sweet-tasting milk,

rich in butterfat and protein. I liken it to drinking half and half. It's not unusual for me to cut it with a bit of water if I just want to enjoy a glass with a few cookies! Mine produce more milk per capita on less feed than either of my standard breeds, so it is an economical breed as well. If forced to choose one breed to raise, the Nigerian Dwarf would win hands down—no hesitation. I love the breed with all of its personality traits, both good and bad, its ability to reproduce with few problems and the fact that it's an easy and economical breed to raise.

"While the attractive coloring and appearance are what originally caught my eye, it's the temperament that has kept me breeding these beautiful creatures for more than twenty years. I often joke that if an Oberhasli makes noise, it's time to call the vet. As a rule, they are a very quiet breed. Even at feeding time when many other breeds become quite vocal, the Oberhasli stand quietly waiting for their rations. They are just too dignified to make a ruckus," Tom explains. He continues

> Rarely have I had an aggressive Oberhasli. Even mature bucks during breeding season are easily handled and are able to be penned with other bucks of varying sizes without harm to anyone. They will do the typical head butting of any goat, but it is usually more posturing than contact and is over very quickly and the combatants are often found minutes later nestled together taking a nap. While genetics and environment are both components of temperament, most Oberhasli love people attention but are not pushy about getting it and have no problem sharing the pats and scratches with their herdmates.
>
> While the popularity of the breed is growing, they are still far less numerous than some of the other breeds. Finding a quality Oberhasli to add to (or start) your herd may require a bit more effort but is well worth it.

Saanen

A solid white or cream colored, large goat, the Saanen originally came from Switzerland. Sometimes called the Holsteins of the dairy goat world, Saanens tend to be excellent milk producers, although the milk is fairly low in butterfat. This makes them a popular choice for goat dairies with a focus on fluid milk rather than cheese. The Saanen has been the top-producing breed in milk production for many years, usually averaging more than 2500 pounds in a standard lactation. Butterfat averages only 3.3 percent, but with such high production, they can be a good choice for someone who wants lots of milk and cheese.

Heather Houlahan of Harmony, Pennsylvania, originally bought two Saanens because they were available in her area, and it turned out to be a good decision for her. "They are tremendous milk producers," she says.

Kelli is a Saanen owned by Dawn Penn at Triple Creek Dairy.

"I can accumulate enough milk for a little cheese making even if only one of them is milking."

Sable

Historically, if a Saanen goat in the United States was born any color other than white or cream, it could not be registered as a Saanen, and hence the Sable Saanen breed was born. The name has now been shortened to Sable, and they can be any color other than solid white or cream. Sables are somewhat rare and are not easy to find in most parts of the United States. Milk production and butterfat are similar to the Saanen.

Toggenburg

The Toggenburg also originated in Switzerland and is the smallest of the standard breeds, with a minimum height of 26 inches for does and 28 inches for bucks. Toggs are only brown, but the shade can vary from light fawn to dark chocolate, and specific white markings are acceptable, such as two white stripes down the face and down the lower part of the legs. They average around 2200 pounds of milk with 3.1 percent butterfat.

Grade or experimental dairy goats

When you are raising more than one breed of dairy goats, odds are good that at some point a buck is going to jump a fence when a doe is in heat, and crossbred goat kids, often called grade or experimental, will be the result. A recorded grade is the offspring of registered parents, and they may not be less expensive than a purebred goat. Some recorded grade goats are on DHI milk test and have distinguished themselves as excellent milkers, and if so, you can expect to pay as much for their kids as you would for a kid from registered purebred parents.

Meat breeds

Boer, Kiko, Tennessee Fainting (or Myotonic), and Spanish goats are meat goat breeds in North America. Although excess bucks are a fact of life with dairy goats, excess milk is not common with meat breeds. Because meat breeds have not been selectively bred for milk production, milkable teats, and good udders, very few meat goats make great milk goats. They also have shorter lactations than you find in better dairy goats. Because they are not bred for milk, there are no official milk records for these breeds.

If you start with a couple of dairy goats and discover that you want more meat, you can increase the size of your herd so that you'll have more kids for meat. You can also add a doe to your herd that is a meat goat and breed her to your dairy goat sire. Although the kids won't be as big as if you were breeding her to a larger buck, they will be bigger than what you were getting when breeding the dairy goats.

Fiber breeds

Although dairy goats should be able to provide you with as much milk, meat, and leather as you want, you might need to add Angora goats to your herd if you want homegrown fiber. All goats produce cashmere as part of the winter coat, but the volume of the cashmere is very small and it must be separated from the coarse guard hair that makes up the majority of the goat's coat. Angora goats, which produce mohair, are smaller than standard-sized dairy goats but larger than the Nigerian Dwarf. Unlike sheep, which are sheared annually, angora goats are sheared twice a year as their long curly locks of mohair grow to several inches in only a few months.

Angora does bred to Nigerian Dwarf bucks produce Nigoras, which are considered a multi-purpose goat that will yield milk, fiber, and meat.

This is still an experimental breed, however, and production numbers are not available.

If you decide to start crossing breeds, remember the buck must be from the smaller breed in order to avoid birthing challenges. An Angora buck can be bred with a standard-sized dairy goat doe because the Angora is smaller in size.

Does

The number of goats you need will be determined by which breed you choose and what you plan to do with the milk and with the kids that are born. The amount of pasture space you have may limit your options. If you need 2 gallons of milk or more per day for consuming as fluid milk, one of the Swiss breeds, such as Alpine or Saanen, would be a good choice. Excellent milkers in those breeds can produce a couple of gallons a day at their peak, gradually declining to around a gallon a day, which they can produce for a few months. Ten months is a standard lactation period, but some does will milk for a couple of years without rebreeding. If you want does that can milk for extended periods, be sure to buy from someone who milks their goats for an extended period of time and keeps records. Keep in mind that first fresheners produce less than mature does and that individual production can vary tremendously between goats of the same breed. For example, the range for Saanens in 2011 was 830 to 6,080 pounds for 285 to 305 days, according to the American Dairy Goat Association. This is why it's important to buy from someone who keeps milk records. Don't assume that a huge udder has a lot of milk in

Official milk records are reported in pounds and tenths of pounds of milk, rather than cups, quarts, or gallons. Weight is used because it is far more accurate than eyeballing a measuring cup or canning jar, and if you are hand milking, odds are good that there will be foam on the milk, making it harder to figure out the exact amount visually. Milk is a little heavier than water, so a gallon of milk will weigh between 8 and 9 pounds, depending on the percentage of milk solids, such as butterfat.

Why Do We Talk About Milk Production in Pounds?

it. While you can't hide a lot of milk in a tiny udder, a goat with a big udder could be a great producer, or she could have a meaty udder.

A breed with high butterfat is the best choice for making a lot of cheese. The high butterfat level increases cheese yield but generally comes at the cost of lower fluid milk production as breeds that have higher butterfat tend to produce less milk. The Nigerian Dwarf has the highest butterfat, averaging around 6.5 percent over the course of lactation, but they average a quart or two a day. Miniature dairy goats, which are a hybrid created by crossing a Nigerian Dwarf buck with a standard

---

You Can't Have Just One…

Remember that goats are herd animals, which means you need to have at least two so that they have a friend who speaks the same language. Keeping them with a pig, sheep, or horse is not the same. Pigs communicate by biting and horses by kicking. Although it may appear that goats and sheep speak a similar language, they don't. Sheep run at each other with their heads down to butt heads and establish dominance, whereas goats rear up on their hind legs and come down to butt heads.

Buying a single goat is asking for trouble. I have received plenty of phone calls from people who bought a single goat and then frantically searched for another one. A lonely goat will be the world's best escape artist because it is looking for a friend. One woman said her goat wanted to live on her front porch, which meant it was hard to keep the poop and pee cleaned off. Another said her goat would jump on her car and dance around. One caller was terribly afraid that her horse was going to kill her goat. In spite of the fact that the horse kept kicking at it, the goat kept sneaking into the horse's pasture for company.

Although there may be solo goat success stories, it isn't worth it to try. Goats do not unlearn bad behavior. Once a goat starts doing something, it will likely teach the trick to its new friend when you bring in another goat. There is simply no reason to buy a single goat. It takes just as much time to care for one goat as three or four, and most reputable goat breeders refuse to sell single goats to homes where there are not already other goat friends.

breed doe, have butterfat that is almost as high as the Nigerians but produce more milk. Nubians have butterfat around 5 percent, although they have the lowest overall production of the standard breeds. Other standard breeds of dairy goats average around 3 percent butterfat, which is similar to whole cow milk sold in the store.

If you plan to use the extra bucklings for meat, the standard dairy goats obviously produce larger kids and therefore more meat, but how much goat meat do you want versus milk or cheese? The Nubian is one of the meatier dairy goats, so if you want a lot of cheese and meat, this breed might be a good choice for you. Saanens, Sables, and Alpines will provide lots of meat and fluid milk. If you want cheese or a smaller amount of milk but no meat, the Nigerian Dwarf might be a better option. Extra Nigerian Dwarf bucklings can often be castrated and sold as pets because of their small size, although they can also be butchered.

Bucks

To make milk, a doe has to get pregnant and give birth, which means you need access to a buck—or at least a buck's semen. For many people this means buying a couple of bucks to breed the does. However, some people don't want to buy a buck or can't have a buck for some reason, such as living in a city and being limited by zoning that allows only two or three does. You might also be worried about the odor of a buck bothering you or your neighbors if you have a small piece of property. There are a few options for those without a buck.

You can buy a bred doe, but that will work for only the first year. After that, a buck can be leased and brought to your property for a month or two. He stays with the does and breeds them when they come into heat. On the flip side, you can wait until you see your doe in heat and take her for a date at the buck's farm. Some breeders will also provide boarding for a doe to stay at the buck's farm for a few weeks if you are having trouble figuring out when she is in heat. However, most breeders do not offer buck service because of biosecurity issues. This is something you can discuss with breeders when buying does from them. Some will breed does that were born on their farm, even though they won't offer buck service to goats that were born elsewhere.

Prospective goat owners may be hesitant to have a buck because they have heard horror stories. Genetics can play a role in personality, so talk to the breeder of a potential herd sire about his sire's personality. While standard-sized bucks can be a challenge to handle, Nigerian Dwarf bucks tend to be mellower. Because they weigh one-third to one-half as much as a standard buck, they are easier to handle.

You might also consider artificial insemination (AI). Although semen costs far less than a buck costs, a tank for storing semen will cost considerably more than all but the most expensive bucks, and you will need to have the tank recharged regularly so the semen stays frozen. You would need either to have an AI technician inseminate your does for you or learn the technique yourself. Success with AI is usually not as good as with a live buck.

Registration

The largest dairy goat registry in the United States is the American Dairy Goat Association, which has registered more than a million goats since it was founded more than one hundred years ago. The American Goat Society and Canadian Goat Society (CGS) are two smaller organizations that also register dairy goats. The registries keep track of goat pedigrees. They also sanction goat shows, license judges, administer classification and appraisal programs, and create criteria for goats to earn milk stars on 1-day and 305-day milk tests.

You might think that you need purebred, registered goats only if you plan to show, but keep in mind that it costs as much to feed and care for an unregistered animal as it does to care for a registered one. There are some distinct benefits to buying registered goats even if you plan to raise goats only for your family's milk and meat needs.

- Unregistered goats may be poor quality. Reputable breeders will not sell an animal with papers if it has a disqualifying defect. Therefore, you could be buying a goat with a defect or a goat whose parent had a defect.
- Kids of registered goats can be sold for more money than kids of unregistered parents.
- Registered goats may have documented show records, evaluations,

and milk tests that will help you determine whether you are buying good quality animals.

- Registered goats have pedigrees so you can see how their parents and grandparents performed in the show ring and in milk testing.

But do you need show quality goats if you are planning to use them only for milk? The short answer is no. Goats from champion parents may not necessarily produce more milk than goats from parents who never set one hoof into a show ring. Great conformation and milk production do not always go hand in hand, although it's great when they do. Given the choice of a doe from a finished champion that would cost the same as two does from a goat with a great milk record, the smarter choice would be to buy the two does from the great milker for the same price. Very few people show their goats, so kids with champion grandparents and great-grandparents do not necessarily sell faster or for a higher price than kids without champions in their pedigree. Although you might be excited to pay top dollar for a goat with champion parents, remember that its kids will be the grandkids of the champion goats, diluting the genetic inheritance. To command top dollar for your kids, the goats that are the parents of those kids have to have proven themselves in either production or the show ring.

If you are planning to milk your goats, you should buy from someone who milks and keeps barn records or is on official milk test, also known as DHI, which stands for Dairy Herd Improvement. Herds on official test will have their milk weighed and tested for butterfat and protein monthly. DHI can be expensive for people with small herds or

What's Show Quality?

The phrase "show quality" does not have any official definition and can mean something different from one person to another. A show quality kid might be defined as one from champion parents, or it might simply mean that it does not have any disqualifying defects. The bottom line is that you can't tell when a kid is two or three months old whether it will grow up to become a champion.

no official tester nearby, but owners can still keep barn records, which means they record the daily weights to track their goats' production. There is no way that anyone can tell with accuracy whether their goats are good milk goats if they do not milk them for extended lactations, which are at least eight to ten months.

Pedigree

The alphabet soup that surrounds a goat's name on a pedigree can be confusing for people new to the goat world. What do all those letters, pluses, and asterisks means? And does it really matter?

Although letters vary between registries for similar achievements, the letters at the front of a goat's name usually signify the goat's championship status and sometimes its milk status. Letters after a goat's name signify milking records and classification or linear appraisal scores. For example, ARMCH Antiquity Oaks Carmen *D VG is a master champion (MCH) with the American Goat Society, and she has an advanced registry (AR) milk star, which means she earned her milk star on a 305-day test. The *D means she earned a milk star, which could have

---

Reading a Pedigree

In both ADGA and AGS, there is a system in place to have a goat evaluated against the ideal standard. In ADGA it is called linear appraisal (LA), and in AGS it is called classification. Although the evaluation and grading systems are not identical, it is highly unlikely that an animal would score extremely well in one and poorly in the other. The letters E, V, VG, G, G+, A, or F or a + sign, after a goat's name indicate the classification or LA score. In both registries a score of 90 percent or more is Excellent (E); a score of 85–89 percent is Very Good (V) in ADGA and (VG) in AGS; 80–84 percent is Good Plus (+) in ADGA and Good (G+) in AGS; a score of 70–79 percent is Acceptable (A) in ADGA and Good (G) in AGS; and 60–69 percent is Fair (F) in both registries.

As a matter of practicality scores of less than 80 percent are not usually advertised because the market for "acceptable" goats is not very big. Those are the goats that go to the sale barn where no one cares about LA scores.

American Dairy Goat Association (ADGA)	American Goat Society (AGS)
DOES	
1*M: A one-star milker is a doe that has met the minimum requirements to earn a milk star, whether in a 1-day or 305-day test. A doe may also earn a star if she has three daughters that have earned stars or two sons who have earned +B.	***D:** A star dam is a doe that has met the minimum requirements to earn a milk star, whether in a 1-day or 305-day test. A doe may also earn a star if she has three daughters that have earned stars or two sons who have earned +S.
2*M: A two-star milker is a second-generation doe that has met the minimum requirements to earn a milk star, whether in a 1-day or 305-day test. A 3*M would be third generation, and so on. Stars cannot skip generations, so if a doe's grandmother is a 1*M but her mother is not, she will be a 1*M rather than a 2*M.	**2*D:** A two-star dam is a doe that is the second generation to meet the minimum requirements to earn a milk star, whether in a 1-day or 305-day test. A 3*D would be the third generation to meet the requirements, and so on. As with ADGA, stars cannot skip generations.
CH: The doe has won three grand champions at shows with at least ten does entered.	**MCH:** The doe has won three grand champions at shows with at least ten does entered.
GCH: The doe has won three grand champions at shows with at least ten does entered and she has earned a milk star.	**ARMCH:** The doe has won three grand champions at shows with at least ten does entered and she has earned a 305-day milk star.
BUCKS	
***B:** A star buck's dam and his sire's dam have earned their milk stars.	***S:** A star buck's dam and his sire's dam have earned their milk stars.
+B: A plus buck has three daughters out of three different dams who have earned milk stars, or two sons who have each earned +B.	**+S:** A plus buck has three daughters out of three different dams who have earned milk stars, or two sons who have each earned their +B.
++B: A two-plus buck has three or more daughters who have earned milk stars and two sons who have each earned +B.	**++S:** A two-plus buck has three daughters who have earned milk stars and two sons who have each earned +S.
CH: The buck has won three grand champions at shows with at least ten bucks entered.	**MCH:** The buck has won three grand champions at shows with at least ten bucks entered.
GCH: The buck has won at least three grand champions at shows with at least ten bucks entered, and he has earned +B.	**ARMCH:** The buck has won at least three grand champions at shows with at least ten bucks entered and he has earned +S with at least three of his daughters earning their 305-day milk stars.

been a 1-day test or 305-day test, and the VG means she scored "very good" when classified. If she had not earned her master championship, which requires at least three show ring wins against at least ten goats at each show, there would not be anything on her registration papers to signify that she had earned a 305-day milk star, rather than a 1-day milk star. Breeders who are on year-round testing usually have milk records on their websites or can give you the numbers so that you can see exactly how much they have milked in the past.

When breeders have goats that are dual-registered, you may see letters from both registries in the animal's name, such as ARMCH Antiquity Oaks Carmen *D 1*M VG. In this case, Carmen has also been on milk test with ADGA and earned her milk star with that registry as well. Registration papers for each registry contain only the recognition earned with that registry. It is only on a farm's website or in advertisements that you will see a name listed with letters from both registries.

When I was still fairly new to goats, I became totally star struck and wanted to buy goats with milk stars in their pedigrees, not realizing that there can be a difference of hundreds of pounds of milk between two goats each of which has a milk star. A goat that barely squeaked by to earn her milk star is definitely not as valuable as one that is on the Top Ten list of breed leaders. And if you want goats to provide milk year-round, a 305-day milk record is definitely more important to review than a 1-day milk star, which is simply verification that a goat milked a certain number of pounds on one day of her lactation. Some goats can produce a lot of milk early in lactation but then dry up after only a few months, so it's important to look at long-term milk records.

Genetics is a gamble, and if you are buying kids, you are buying genetic potential based on the goats in that kid's pedigree. Although you can't be guaranteed a bucket-busting milker based on a great pedigree, it is highly unlikely that you'll get a great milker from a mediocre dam and grandmother. On the flip side, don't get too excited about goats that are far back in a pedigree. Having an outstanding milker as a great-grandmother in a goat's pedigree only represents one-eighth of that goat's genetics, and if the rest of the goats in the pedigree are only mediocre, the odds are against the kid getting only the spectacular genetics.

Purchasing

Although you can find goats on Internet classified ad sites and in sale barns, the quality is often questionable. Keep in mind that no one is going to sell a goat for $50 if they can sell it for $300 or more, and if they can sell it only for $50, there is probably a very good reason. The animal could have a disease, a disqualifying defect, kidding problems, or poor milk production.

An Internet search will find the websites of breeders in your area. Search for your state and the breed you want, such as "Oregon Nubians." Many goat breeders have websites where you can learn more about the herd, the breeding philosophy, and their individual goats. They often have photographs, milk records, show records, classification or linear appraisal scores, and sometimes even stories about the goats.

It is a good idea for a couple of reasons to buy your first goats from a breeder who has a philosophy similar to yours. First, if the goats are thriving in their current environment, they may not perform as well under a different management system. For example, if you want to raise your goats in a sustainable system, you won't know if goats from another farm will do well on your farm if they are being given multiple vaccines and chemical dewormers on a regular basis. Also, if a farm bottle raises all of their kids, they will have no idea whether their goats are good mothers, which could present a challenge if you are planning to dam raise. Hopefully the person who sells you your first goats will be willing to serve as a mentor for you. Having someone who shares your philosophy and personally knows your goats is invaluable.

Keep in mind that if you want good quality stock of a specific breed, you might need to look at surrounding states or even across the country. When I was starting my herd, there was no one in my state who was raising Nigerians for anything other than pets, so my foundation animals all came from other states, including bucks that came from as far away as Massachusetts and Alaska. Goats can be shipped by air in dog crates. When buying from a distance, it is less expensive to buy kids than adults because the cost of shipping is based on weight or size of the crate.

Should you buy all your goats from a single herd? From a veterinary perspective, it is safer to buy all of your goats initially from a single herd.

If you are bringing in animals from a variety of places, they will each come with their own germs and parasites. Although each individual goat has been living with its bugs forever, the other goats have not, and their systems will be faced with the stress of fighting off new bacteria and viruses. It is impossible for a goat to be completely parasite free, and with the growing problem of dewormer resistance, bringing in goats from a variety of herds could result in severe parasite problems. There are only three classes of dewormers, so if you bring in goats from three different herds, it is possible that you will have put together goats that are carrying parasites that combined have resistance to all known dewormers.

Probably at some point you will be bringing in at least a few goats from different herds, and when you do, it is imperative that you quarantine new goats for both their safety and the safety of the rest of your herd. Moving to a new farm is stressful on goats, whether it is across the road or across the continent, and being isolated from other goats can stress them even more. If you can afford to do it, bring in two goats at a time from a single farm so that they will have a friend to stay with them during the quarantine period.

If you buy a single goat, put a wether with it for company to keep its stress as low as possible. Although it may appear that a castrated male would have no place on a farm, he can provide a variety of important functions, such as letting you know when does are in heat as well as being a companion to animals in quarantine, such as those newly purchased or showing signs of illness. They can also help out around a homestead by pulling a cart or carrying firewood. Because they are not producing babies or milk or sperm, they tend to be very easy keepers with high resistance to parasites and illness.

How do you know you are buying healthy goats? Goats can have a number of health problems, and some are more obvious than others. Few people would be willing to take home a goat with a crusty nose or diarrhea, but there are some diseases that can be asymptomatic in the early stages.

Caprine arthritis encephalitis, usually called CAE, and Johnes (pronounced like yo-nees) can often be detected only by testing animals. There is no requirement for testing, and everyone handles it a little differently. Some herds are tested annually, especially if they attend shows and the goats are exposed to other herds. A herd may be closed, meaning the breeder no longer buys goats and does not offer breeding services. But a herd may be called closed even though the animals are taken to shows, so the term "closed herd" does not have a universal definition. After several years of negative test results in a closed herd that does not show, some breeders may test less often or not at all.

If the animals you want to buy are not tested, you can ask that adult goats you want to buy be tested or the dam of any dam-raised kids you want to buy be tested. Because it takes months for a goat's body to develop enough antibodies to show up in a blood test, pathologists recommend that kids not be tested until six months after they have stopped nursing. As an additional precaution, you can ask to have the kid's sire tested because research has shown that CAE can be transmitted through breeding.[1] If the dam was infected during breeding she could infect nursing kids with the virus, even though she might still test negative. Normally the buyer pays for pre-sale testing. Although a single negative test is not as convincing as several years of whole herd negative tests, it is better than nothing. More information about CAE and Johnes is in the health chapter.

One should be concerned about tuberculosis (TB) and brucellosis when buying dairy animals because these diseases can be transmitted to humans through body fluids, such as blood, milk, and vaginal secretions during birth. Tests are available for both of these diseases, but the incidence of TB and brucellosis in humans is quite rare because of aggressive programs to eradicate the diseases in dairy herds. According to the Centers for Disease Control, around one hundred cases of brucellosis occur annually in the United States.[2] Almost all states are accredited tuberculosis-free, and many have not had a case of TB in twenty-five years, but this can change literally overnight if a new case is discovered. This is why most TB-free states have strict rules about importing animals from states that are not accredited TB-free.

When animals cross state lines, whether in an airplane or private vehicle, they are supposed to have a certificate of veterinary inspection, often called a health certificate. In many cases it simply contains information on the seller and the buyer, along with goat identifying information, such as tattoo numbers. A veterinarian signs the health certificate, signifying that the animal is not exhibiting any signs of disease. When a goat is coming from a state with a known disease problem, such as TB, the form will also include the test results required by the state the animal is being imported to.

What I learned from Tom Selleck (the goat, not the actor)

After two years of goat ownership, I bought Tom Selleck, a new buckling, and immediately put him into the pen with my other bucks. A couple weeks later I took a fecal sample to the vet. She said he had a heavy load of barber pole worms and tapeworms and should be given a dewormer for three days. A week later we found him unable to stand and rushed him to the University of Illinois veterinary hospital. Within less than an hour of our arrival, he was dead.

A necropsy showed that he had died from anemia caused by the barber pole worms. I was confused because we had just given him a dewormer. When I talked to the woman at the farm he had come from, she said, "Well, everyone knows that dewormer doesn't work." That was my first lesson in dewormer resistance. The particular dewormer had always worked well on my farm, but the internal parasites on the other farm had developed resistance to it. Not only did I lose a buckling, but we then had worms on our pasture that were resistant to the dewormer we had been using.

The vet told me to start deworming the bucks monthly, which I did, but a couple months later, two more bucks died from parasites. It was bad enough to lose one goat, but it was even worse to ultimately lose three. Although quarantining might not have saved the buckling I had purchased, I would not have lost the other two bucks.

You may think that if you are buying your goats from a herd that has tested negative for all of the most insidious diseases there is no need to quarantine. However, it is sometimes the simplest things, such as parasites, that can cause the biggest problems.

CHAPTER 2

Housing Your Goats

✦ ✦ ✦

ONE OF THE ADVANTAGES that goats have over cattle, sheep, and pigs is that the equipment and infrastructure required for them is not nearly as costly. Goats were the first livestock we bought after chickens, and we had to buy very little equipment for them. When we added other livestock, however, we quickly learned that we would need to upgrade our infrastructure to properly contain them. Cattle require heavy-duty steel handling equipment, which is very costly. Although ewes are not any harder to keep than goats, the rams are very hard on housing and gates. The first time we put a ram in our one of buck pens, he put his head down, ran straight for the gate and busted right through it. Pigs also tend to tear up a lot of buildings and fencing simply by rubbing on it. You will probably discover that goats are easier to keep than you imagined.

Shelter

Many people who live in northern climates assume they will need insulated and heated barns for their goats in the winter. However, goats grow a thick, fuzzy undercoat of cashmere to keep them warm during the winter, so adults are usually fine in unheated barns in most of North America. If kids are born in freezing temperatures, someone should be there to get them dried off as quickly as possible so they don't get

A three-sided shelter works well for buck housing. Just be sure that the opening is on the opposite side of the building from the prevailing winds during the winter. For us, the openings are on the south side. However, when we had a blizzard a few years ago, we brought the bucks into the barn, and we were very happy we did because the three-sided shelters filled up with snow. If you don't have a backup housing plan for bucks, use bales of straw in front of the opening in case of a blizzard.

hypothermia and their ears don't get frostbitten while still wet. Once kids are dry, they are fine down to around zero Fahrenheit.

Goats need to be protected from snow, rain, and wind because these things will cause a lot more stress than cold temperatures alone. Unlike sheep and cattle, goats seem to think they will melt if they get wet, so most of them will start to scream wildly if it starts raining when they are outside. Three-sided shelters in the pasture are ideal because the goats can get out of the rain and wind, but they still have fresh air.

A barn with big, heavy doors that you have to open and close to allow goats to go in and out can be modified by cutting a smaller, goat-sized door into the side of the barn so that the goats can come and go as they choose when there is threat of rain or other inclement weather. In general, however, the goats should be outside during the day, where they can get fresh air.

You may have heard of using an old doghouse as a goat shelter, and although this can provide shelter, it may not work for more than one goat. There is a hierarchy in every herd, and if the shelter is not big enough, the less dominant goats may be left outside in the rain. If you

We have cut several goat doors in our big barns to make it easier to put goats out to pasture. (Photo courtesy of Sarah Schwimmer.)

This is the inside of a goat barn that will accommodate up to five Nigerian Dwarves, three miniature or two standard-sized dairy goats. The building is 12 feet by 12 feet and is split into two sections. The goat area is 7 feet wide and has a dirt floor. There is a goat door that leads to a fenced area. The area behind the half wall on the left has a floor and is used for hay storage and milking. (Photo courtesy of Lyn Adams, New Salem, Massachusetts.)

are planning to milk your goats, you will want a building that not only will shelter the goats but will also provide a comfortable environment for humans during birthing and milking.

Although I have heard of people milking their goats outside, it is more comfortable to have a milking parlor to keep you and the goat out of the weather. It needs to be separate from where the goats live; otherwise, all of the goats will be fighting to get on the stanchion and eat the grain in the bowl.

You also need to consider your own comfort when thinking about kidding season and housing. For six or seven years, we sat in the big, cold barn waiting for does to give birth, with other does walking around in the same stall. Finally it occurred to me to use our other barn, which had never been used for livestock and which had a small heated room with windows looking out into the open part of the barn. If you are building a barn from scratch, this is something to consider. I probably would not have thought about this configuration if we had not already had the building on our farm.

We decided to build kidding pens in the barn with the heated viewing room so that we could watch the does without freezing. Kidding pens

Yodel, a young Nigerian Dwarf, shares his home with one other goat. This little house was created for two pet goats, but it could be used for a couple of standard milkers or three or four Nigerian Dwarf does. There is no storage space or place to milk with a set-up like this, so you might milk in a garage, back porch, or outbuilding on the property. This could be used as a buck shelter if the size of the herd increased. (Photo courtesy of Jill Armstrong, Marshfield, Massachusetts.)

provide space for does to give birth in semi-privacy. Our "semi-private kidding suites" have pig panels between them so that the does can see each other but won't be able to butt heads or bother each other when one is in labor. Goats can be very mean to each other, and you never know who is going to be left alone when she is in labor and who is going to be picked on. Being herd animals, they tend to get upset when separated from other goats, so using pig panels to separate them meets their need for company but protects them at the same time.

Bedding

The two most popular options for bedding are straw and wood chips. The purpose of bedding is to soak up urine, cover up poop, and insulate the goats from the cold ground in winter. Both types of bedding do a good job of soaking up urine and covering up poop. However, straw is definitely warmer in the middle of winter. Straw is also a better option during kidding season because wood chips are small enough that they can wind up in a kid's mouth or nose, causing choking or suffocation. Straw is the least expensive option in our area, and we need only one bale of straw compared to three bags of shavings to bed one of our 10-foot by 15-foot stalls. In our area using wood shavings to bed a single stall will cost ten times more than to bed the stall with straw, but this could be different in other areas, so be sure to check prices. If you have a place to store loose wood chips, you might be able to get a good deal on a truckload.

CHAPTER 3

Protecting Your Goats

✤   ✤   ✤

LIKE MOST ANIMALS, goats need plenty of fresh air and sunshine to stay healthy and perform at their peak, which means you can't keep them in the barn all day. It is difficult to say exactly how many goats can be kept on a certain number of acres because it varies widely depending upon your climate, what grows on your property, and the size of the goats. As you learn more about the needs of goats, you will start to get a feel for how many you will be able to keep on your property.

It is impossible to talk about pastures without talking about fencing. People often think that they will need miles of expensive permanent fencing and many acres for keeping goats, but there are alternatives that make it easy for people with only an acre or two to keep goats as easily as those with a hundred acres.

Fencing

The selection of fencing is one of the most important decisions you will make, and hopefully that decision is made before the goats are on your property. If you don't have adequate fencing, you will find your goats visiting the neighbors, eating your rose bushes, and finding all sorts of mischief. The saying "If a fence can hold water, it can hold a goat" is

What I learned from Bucky

I started my herd with three does and a buck. Knowing that goats are herd animals and need company, I was buying my buck from a breeder who was going to give me a wether for company. However, after driving six hours to pick them up, I discovered that the wether had a poopy back end, which meant coccidiosis, and being new to goats, I didn't want to bring home a sick animal. So, I brought home the single buckling, wondering how I would keep him from getting lonely.

We thought we had a great idea. We put together four livestock panels to make a 16-foot by 16-foot square pen for him in the middle of the pasture where the does lived. We had no idea he would be unhappy having a fence between him and other goats, and we spent weeks coming up with one alteration after another to keep him separated from the does. He quickly realized he was small enough to squeeze through the squares in the panels, so we wrapped the whole pen with chicken wire. He learned to hoist himself up high enough to go through the squares that were above the chicken wire. We added more chicken wire. He started jumping over the panels. We bought more panels and put them on top of the pen so that he couldn't jump out. It was so complicated that it took us five minutes to get into the pen, but finally Bucky could not get out!

The following year when we bought a second buck, we built a separate pen for them, and for some reason that I no longer remember, my husband put the latch on the inside of the pen. It was a slide bolt style that you lift and slide to open. One day I looked out into the pasture and saw Bucky chasing after the does. I accused my children of leaving the gate on his pen open and told them to go lock him up. Less than fifteen minutes later he was running around out there again. My children insisted they had locked the gate. This time I went out and put him back in his pen. I turned my back but had not even left the pasture when I realized he was running past me to get to the does. He had learned to open the latch!

Goats are incredibly intelligent animals and will copy human behavior to do things like unlatch gates, turn on lights, and even attempt to open doors. We once had a LaMancha doe that would put her mouth over a doorknob and turn her head in an attempt to turn the doorknob as she had seen us do. If only her mouth had not been wet and slippery, I'm sure she could have actually opened the door!

only a slight exaggeration. But the fact is that without the proper type of fencing, any species of animal is impossible to keep fenced in. What works for goats doesn't always work for cows or pigs.

Even if you already have goats and your fencing seems to be working, don't skip this section. Some people just get lucky with their first goats and don't have any problems with fencing that will suddenly be completely worthless when they have goats that are more motivated.

Chain link

One of the most goat-proof fencing options available is 6-foot chain link. Unfortunately, it is also the most costly. Standard-sized bucks can go over 5-foot fences, and Nigerian Dwarf bucks can go over 4-foot fences, so you really do need the height to keep bucks on the right side of the fence. Because of the cost, very few people use chain link for fencing goats other than bucks, and few use it for that.

Electric—permanent

I cringe when I see articles and books written with the blanket statement that electric fencing works for goats. It is far from simple to use electric fencing with goats successfully. In addition to training goats to electric fencing, you also need to install electric fencing differently for goats than for other livestock, such as cattle, horses, or pigs. Permanent electric fencing is generally individual horizontal wires that are attached to a charger that sends an electric current through the wire to shock any animal (or human) that touches it. You can get expensive high tensile or the less expensive electric fencing that is often sold at local farm supply stores, and neither one will work unless the strands are close enough together, close enough to the ground, and high enough, and the charger is strong enough.

Goats have small feet and are not very heavy, meaning that they are not well grounded, so they won't feel a shock as much as a cow or a human will. They can jump through fencing wires that are not close enough together, and they can jump over fences that are not tall enough. They will even go under an electric fence if it isn't close enough to the ground. I bought my first goats from someone who had successfully used electric fencing for years with her goats, but it didn't work for us. Why? We had moved to a former horse farm, and the three fence wires were a foot apart and a foot from the ground. We thought the goats looked like they were too big to go under it or through it without getting shocked. But the goats thought differently. Although they did get shocked, they really didn't care. By the time we started adding more wires and lowering the bottom wire, it was too late. The goats had gotten into the habit of going through the fence, and they never stopped, even after we had six strands of electric wire, 6 inches apart. We also have hilly pastures, which makes it nearly impossible to get the bottom wire close enough to the ground in every area without having it touch the ground in other places, causing it to short out.

If you do want to use electric fencing, be sure to use five or six strands. Space the strands 6 to 8 inches apart, depending on the size of the goats that will be in that pasture. The charger is also very important. Buying the cheapest charger you can find will probably yield disappointing results.

Using temporary or permanent electric fencing in an urban area is a bad idea due to liability issues. Although there has only been one death ever associated with an electric fence, it is possible for someone to be hurt by the electric shock from the fencing.

Electric—temporary

Temporary electric fencing has a number of benefits over permanent electric fencing. There are now temporary electric fences available for every species of livestock, including poultry, and even for keeping predators out of gardens. The fences are all different, designed for a specific purpose. It is important to get the fencing made for goats. The temporary fencing for goats is made of electrified horizontal wires a few inches apart and non-electric, vertical plastic struts or strings. We have used only the fencing with the plastic struts, which I recommend because the one with vertical strings has more issues with sagging, making it easier for goats to jump over it. Kids are also more likely to get tangled in the netting that is made with vertical strings. Regardless of which style of electric fence you use, temporary or permanent, it is always a good idea to have someone present when kids are first introduced to electric fencing. You

Temporary electric fencing works well for most goats because the electric wires are spaced so closely together.

may need to turn off the fence to pull a kid out of it or to grab it on the outside if it manages to go through as it is being shocked.

Temporary electric fencing is great for a farmette, where goats need to be kept on a small area and moved frequently, but it also works well for multi-acre properties using rotational grazing, which is beneficial for pasture management as well as for controlling internal parasites in goats.

Given enough motivation, my bucks have jumped over the 35-inch high temporary electric fencing. A manufacturer has now come out with a 42-inch fence. It is always a good idea to have two fences between bucks and does to avoid the possibility of unintended breedings, especially when you know a doe is in heat.

Livestock, hog, and cattle panels

People often use the terms "hog panel" and "cattle panel" interchangeably, although they are two different things, and livestock panels are something else. Hog panels are short and have closely spaced horizontal wires, whereas cattle panels are much taller and have more widely spaced horizontals. Hogs don't jump, but goats do, making hog panels completely

Even though a pig panel is taller than a Nigerian Dwarf, I have a couple of talented does that can jump over them if they are motivated. My LaMancha does could clear a pig panel with ease.

worthless for bucks and even some does. You can use them for kids until they get old enough or wise enough to realize they can jump over them.

Cattle panels are tall enough to keep does where you want them, but some standard-sized bucks can jump or climb over them. Many Nigerian Dwarf bucks can even jump them until they are around a year old, when many of them become too heavy or lose enough spring in their hooves that they can no longer jump that high. The spacing between the horizontals is wide enough that kids can squeeze through if they are motivated.

Livestock panels, sometimes called combination panels, are as tall as cattle panels, but they have horizontals, similar to the pig panels, that are closer together near the ground, designed to keep babies from going through. All of my buck pens are made of combination livestock panels, which keep the bucks where I want them almost all the time and keep young doelings from going to visit the bucks when they are still

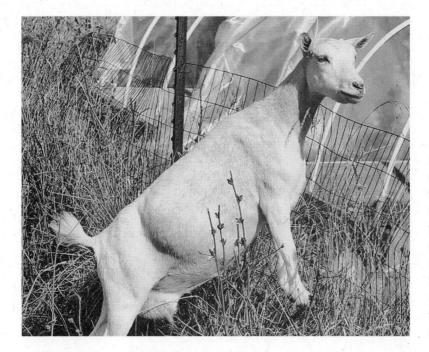

Because goats like to put their front hooves on fencing, welded wire fences tend to fall apart within only a couple of years. Note that this particular fence is not tall enough to keep this doe from going into the garden. A few minutes after this photo was taken, she jumped the fence.

too young. However, this type of fencing does not work for weaning Nigerian Dwarf bucklings. If their dam is on the other side of the fence, most will be motivated enough and smart enough to figure out how to get back to mom.

Welded wire

Although you can get welded wire tall enough to keep goats fenced in, it is one of the more expensive fencing options because the wire is very closely spaced, meaning it uses a lot of metal. You don't need any special equipment to put up welded wire fencing, but the rolls can be heavy, making them a challenge to handle by yourself. The welds do not last a terribly long time, especially if goats rub against them, so welded wire fencing will need to be replaced sooner than other types of fencing.

Woven wire

Most of the fencing for our Nigerian Dwarf herd is made of woven wire. It is less expensive than welded wire and does not need to be replaced as often because the wires are wrapped around each other rather than

Goats are convinced that the grass is always greener on the other side of the fence, so unless you pair woven wire with electric, you will probably find them sticking their heads through the fence to eat the grass on the other side.

welded together. However, a fence stretcher is required to install this fencing. If you are going to use this fencing with bucks and it will be the only fence between bucks and does, you will need to have a single strand of electric fencing running along the fence about a foot off the ground to keep the buck from trying to mate does through the fence. Woven wire is quite stretchy, and I once saw a buck almost succeed in mating a doe that was standing on the other side of the fence. His head and two front hooves were sticking through the fence, and his body had stretched out the fencing over the doe's back. Although my buck was unsuccessful, I know people who have had a buck succeed in mating through the fence.

Woven wire needs to be paired with electric wire to keep standard-sized dairy goats contained. A strand about a foot off the ground as well as one a few inches above the top of the tallest woven wire should do the trick. However, there are no guarantees. I used to have a LaMancha buck that would jump up and mash down a woven wire fence, even when there was an electric strand over the top. He had started smashing the fence before we put the electric wire above it, and he was willing to put up with the shock after the electric was added. If the electric wire is installed with the woven wire, the goats will check it out and probably get shocked, which will cause them to leave the fence alone in the future. Had our buck not known that he was capable of smashing a fence, he probably would not have continued doing so when we added the electric.

Wood

Many of us have bucolic visions of a split rail fence on our homesteads. Unfortunately, this type of fence is pretty much worthless with goats. The rails would need to be extremely close together to prevent kids from escaping, but because the rails are thick, placing them close together creates a fence that is easy for a goat to climb over. Although a 6-foot wood privacy fence would work well, it is a very expensive option.

Regardless of which type of fencing you use for your bucks, be sure that you do not have anything close to the fence that the buck can use as a launching pad to get himself over the fence. Years ago, I heard of a man who was trying to figure out how his buck was getting out of an area fenced with 6-foot chain link when one day he saw the buck get a running start, jump on his shelter, and then catapult himself over the fence.

When Bucks Fly

Livestock Guardians

Good fences are absolutely essential not only to keep your goats where you want them but also to keep predators away. However, even the best fence may prove inadequate when a predator gets desperate enough, which is why many people have a livestock guardian. One or more dogs, llamas, or donkeys typically fill the role of guardian.

Dogs

Although dogs may be the first species most of us consider as a livestock guardian and they can prove the most useful, they can also be the most troublesome. As "man's best friend" dogs such as the Great Pyrenees in France and the Anatolian Shepherd in Turkey have been guarding livestock for centuries. In addition to fighting off larger predators such as bears and mountain lions, dogs scare off or kill smaller predators such as raccoons, which probably won't hurt your goats but are always looking for a tasty chicken dinner. Dogs can also serve as a deterrent to humans who may want to steal your livestock or come into your pasture when you are not around. Many goat breeders have heart-warming stories about a livestock guardian dog that saved their animals from a predator or cleaned up goat kids that were born unexpectedly in a pasture.

Our own Anatolian Shepherd let us know when one of our bucks was very sick, and we were able to nurse him back to health. If the dog had not been standing next to the buck's shelter barking at us, we probably would not have found the goat until it was too late.

"I had an Anatolian named Jazz that knew when the does were in labor before the does did," recalls Penny Oldfather of Illinois. "She would lie or sit next to a doe and let you know she was next. She cleaned off many kids and kept them warm."

A livestock guardian dog should always be introduced to new stock while it is on a lead. There is no way of predicting how the dog will respond, so you want to be able to maintain control at all times.

"I was babysitting a pair of heavily pregnant goats due to kid in some seriously cold weather in an open three-sided shed," recalls Shelene Costello. "Temps were around zero, windchills around 30 below. I just went out and checked them and came back in to sleep a bit. Puppy Artic started to raise cain. I went out to see what the ruckus was, and he was at the fence staring into the goat pen. Sure enough, cold wet babies. But in time to prevent frostbite to any of them. Good puppy. Now as a mature male, he, along with his younger sister and best working bud, spend their days and nights guarding my stock. One goes out with the poultry when they go out to the fields to graze and one stays home with the rest of the stock. They have killed lots of smaller varmints and run off the coyotes, bobcats as well as winged predators."

Akbash	Komondor	
Anatolian Shepherd	Kuvasz	
Great Pyrenees	Maremma	

Common Livestock Guardian Dog Breeds

Unfortunately, there are also stories of dogs injuring and killing goats. I was told about an inexpensive German Shepherd mix that was bought to be put with the goats on a farm. Before long the dog was chewing on kids' ears and had even bitten off a couple, so the family put him in a pasture with only adult goats. One day they found him eating a kid that had somehow slipped into the pasture where he was tied. Stories like this are, unfortunately, too common, as are stories of the beloved family pet that didn't understand it was not supposed to "play" with the goats.

Although we always had rescue dogs as pets, we had to learn the hard way that livestock guardians are very different from pets. They have a job to do protecting the animals on your farm, and just as you would not hire a completely incompetent person to do a job, you should not get a dog to work as a livestock guardian unless the breed has centuries of livestock experience behind it. It is also a good idea to buy a dog from a farm where it has been living with livestock since the day it was born.

Do not confuse herding dogs, such as border collies, with guardian breeds. Herding dogs have a non-stop drive to chase animals, which is completely different from the normal behavior of a guardian breed. Although some breeders say that English Shepherds are dual purpose and can be good guardians, it is primarily a herding breed, and many are far too active to be left alone with livestock until they are older.

Llamas

If coyotes are the main predator in your area, you might consider a llama or two as guardians. An advantage of llamas is that they eat the same thing as goats, so feeding them is not a separate chore, as it is with a dog, and you don't have to worry about the goats eating the dog's food. You will, however, have to have the llama sheared annually so that it doesn't overheat in its long coat over the summer.

An intact male should never be used as a guardian for goats because it may try to mate does when they are in heat. Because the male llama lies on top of the female during mating, it will wind up crushing a female goat.

Llamas protect livestock from coyotes through a variety of techniques. Unlike most prey animals, llamas tend to be more curious than cautious. Rather than running away from strange things, they often run towards them. And because coyotes are small and rather cowardly, they will generally run away when they see something 6-feet-tall running towards them. If a coyote does decide to engage a llama, the llama will stomp and bite the coyote. Llamas have fighting teeth, which are four canines that can rip and tear flesh. Fighting teeth are generally removed on male llamas because of the damage they can do to each other, but these teeth are left on females because they fight only when being attacked.

We added llamas to our farm after we lost almost all of our lambs to coyotes one year. We know the llamas stopped an attack on a ram and another on a turkey hen. Considering how much our losses decreased after adding the llamas, we feel certain they thwarted the attempts of many coyotes.

We once found a female llama in the middle of a frozen creek, unable to stand up because of the slippery ice. There were canine paw prints around her and blood in the snow, but we couldn't find any injuries on her. We can only assume that a coyote thought he had found an easy meal when he spotted her down on the ice, but she did a good job of protecting herself.

I have occasionally heard of someone selling alpacas as livestock guardians, but alpacas are generally too small and timid to be effective. In fact, llamas often serve as guardians for alpacas.

Donkeys

Donkeys are another option for guarding goats if coyotes or dogs are the main predators. Like llamas, they have a diet similar to goats. However, coccidiostats can be fatal if consumed by equines, so goat minerals or a feed that contains a coccidia preventive need to be placed in an area that is inaccessible to the donkey.

Donkeys will also stomp and bite coyotes, and although they don't have canine teeth, they can do a lot of damage when their strong jaws clamp down on a coyote's leg. In fact, we once had a donkey that bit my daughter and left a 4-inch bruise across her shoulder blade.

PART II

Raising, Remedies, and Reproduction

Although some people would classify the information in this section as "management," I like to think of it as working with my goats. Some might say that the difference between the two is simply semantic, but I have seen a shift in my own "management" style over the years. Managing implies the goat keeper has control, and you will never have complete control over living things. There are simply too many variables and too many unknowns, and you will be disappointed as long as you think you can control everything. Although some management practices work some of the time, you have to understand goats and their needs to get the best results.

Working with your goats requires a different attitude. Hopefully, by the end of this section it will become clear that management practices must be different from farm to farm and even goat to goat because every farm and every goat is different. Working with my goats is different from working with any other herd because no one else has the same combination of animal genetics, well water, and availability of hay—important variables that will be fully explained in this section.

Working with my goats also means I give them most of the credit whenever anyone compliments our cheese, yogurt, or soap. Without their hard work giving birth annually and producing milk for ten months of every year, I wouldn't have the basic ingredient to make all of those products.

CHAPTER 4

Day-to-Day Life With Goats

✤ ✤ ✤

Living with goats is a continual adventure in animal husbandry, psychology, and the culinary arts. Every goat is unique, and as my daughter once said, even if we owned goats for a hundred years, we would still be learning. Each goat's milk is different, and each has a different immune system and a different way of giving birth. Personalities are also different not only from one goat to another but depending on whether the goat is interacting with her own kids, other goats in the herd, or humans. One of the most common questions asked by new goat owners is, "Is that normal?" This section will attempt to explain "normal" as much as possible. As you read it, however, keep in mind that there is great variation in what constitutes normal from one goat to another.

Behavior

Goats have been domesticated for thousands of years, but as with most domesticated animals, the degree of friendliness will depend on how much they are handled as babies. Just as cats can revert to a feral state in one generation if born in a place where they get no human interaction, goats will also be wild if not handled as babies. Some people think that goats need to be bottle-fed to be friendly, but the reality is that they simply need attention, and bottle-feeding certainly means that kids will get attention every day. Basically, the more kids are handled, whether you give them a bottle three times a day or sit in the straw and play with them three times a day, the less fear they will have of humans. And some goats simply will be friendlier than others from the beginning. I noticed early in my goat keeping that friendlier mothers had friendlier babies, and more ornery mothers had more skittish babies.

There is definitely a genetic component to personality, and in fact, there is a genetic test for cattle that tells you the genetic predisposition to being calm and to producing calm offspring. According to Igenity, a company that does genetic testing of cattle, docility is an important trait for the beef industry because calm cattle eat better, gain weight better, are healthier, and have more tender meat than those that are nervous

What I learned from Coco

In our second kidding season, we wound up with our first two bottle-fed kids. The first one had hypothermia and was too weak to stand or nurse, and the second one nearly starved to death when she was two weeks old because she had difficulty competing with her three brothers for the dam's two teats. I initially called her PeeWee because she was so tiny, but after a couple of weeks, I realized she was going to survive and gave her a real name—Coco Chanel. Her mother was a difficult goat to handle and did not like being milked. I sold her when Coco was a couple of years old. Although Coco was bottle-fed and is incredibly friendly, she can be incredibly stubborn and quickly taught us that bottle-feeding a kid does not guarantee a goat that is always easy to handle.

My daughters enjoyed showing goats in the early years, and Coco was often the most difficult goat to show. She would plant her feet as if they were stuck in concrete and refuse to move! Other times she would prance around the show ring like a diva! As I write this, she is nine years old, and even today, it can be a challenge to lead her from one place to another most of the time. However, if I go out to the barn or pasture and sit on the ground, Coco wants to crawl into my lap just like she did so many times when she was a baby.

and less calm.[3] I have noticed even in my goats that some of the adults that were dam raised are easier to handle than some that were bottle-fed as babies.

Social order

One of the hardest things for humans to understand is why goats are so mean to each other. And sometimes the goats that are the most outgoing towards humans are the roughest with other goats. They butt heads and slam their heads into other goats' sides. It can be scary to those of us looking on and tends to be especially bad when a new goat is introduced to the herd. For this reason, I try very hard never to add an individual goat to the herd. When goats have been separated for kidding, I put them into groups of three or four initially so the head butting gets spread around a little more than if only two goats were put together.

It is also important to be sure that you have plenty of space for all the goats to fit in front of hay feeders and feed pans. If the goats are crowded, the more dominant goats will get most of the feed. When feeding grain in a pan, it is best to use a fence-line feeder rather than a pan in the middle

A feeder attached to a fence or a wall discourages head butting during feeding because all of the goats are facing the same direction.

of the pasture because goats will butt heads over the pan and often wind up running through the pan, knocking it over, and spilling the feed. If you already have feed pans, placing them next to a wall or in a corner will reduce head butting compared with placing them in a space where the goats can circle around the pans.

Usually, in spite of the severity of the head banging, no one gets hurt. Every now and again, a goat might wind up with a little blood on the top of its head, especially if it has scurs, but long-term injury is extremely rare. In most cases one of the goats will give up and refuse to continue fighting, but I did have a buck wind up with a concussion once. I had owned goats for eight years when one night I saw two bucks butting heads at sundown. I ignored it because no one had ever been hurt in the past. The next morning, however, the smaller buck was staggering around and stumbling, and his eyes were operating independently of each other, moving in different directions. Luckily, he did recover, but I no longer ignore bucks fighting.

Even scarier than bucks butting heads, however, are pregnant does fighting. In most cases goats will butt heads for a few minutes when first introduced, and once in a while they'll hit each other with their heads if they want hay or grain that the other one is eating. But every year or two, there seems to be a doe that simply no one likes. They won't let her have her share of hay, and sometimes they'll even try to keep her out of the shelter. It can get especially scary if the underdog is pregnant. In those cases it's a good idea to put her in a different pen with a younger doe as a companion.

Intervention

Do not ever put yourself between two goats that are fighting, not even if they are little Nigerian Dwarf does, because they can hurt you. Those hard little heads can pack quite a wallop against your knees or shins. A collar on a goat makes it easier to handle the animal whenever you want to move it from one place to another. To separate two goats that are seriously trying to hurt each other—not just a few head butts—stand behind one and grab it by the collar and pull it away from the other one. Do not turn your back on the other goat, especially if it is a buck. If you are dealing with 250-pound standard bucks, you need someone else to handle the other buck. The fact that you are holding one of the goats will not stop the other one from taking advantage and slamming into the goat that you are holding. When you separate goats, especially bucks, you need more than just a fence between them because they can continue butting heads through a fence and wind up hurting themselves. One of my bucks injured his eye butting heads with a buck in the next pen. In most herds you won't have to separate goats very often. With a herd of

Collars

While cattle and sheep usually wear halters, dairy goats usually wear collars. Most goats are taught to lead when someone is holding the collar, and with standard-sized goats, the collar is right at the level of your hand, so it's very convenient and comfortable. With Nigerian Dwarf or miniature dairy goats, however, you may have to bend over a little. For this reason, some people will use a leash on a smaller goat when leading it a longer distance. You can use simple dog collars, or you can buy special plastic link collars that break under pressure so that your goats don't accidentally hang themselves on a tree limb when browsing.

When a goat is browsing, it can easily get its collar caught on a tree limb, which is why some people use break-away collars.

around twenty does and five bucks, I don't think I've ever had to do it more than once a year. And after a few hours, tempers have cooled, and the goats can be reunited again. However, if I find that the herd never really accepts a goat, I will sell that animal because quite simply, I feel sorry for it. I've only had to do this twice, and in both cases it was young does that had been born here and were continually picked on from about a year old until I finally sold them a few months later.

Anatomy

Unlike the majority of mammals, including humans, dogs, cats, rabbits, horses, and pigs, which have only one stomach, goats have four stomachs, which means they fall into the category of ruminants, which also includes cows and sheep. Ruminants spend a huge part of the day eating and re-eating their food. A goat walks around in the pasture eating grass, weeds, tree leaves, and small bushes, and when the rumen gets full, it lies down somewhere and starts the process of bringing up its cud to chew everything a second time and send it on to the next stomach. The rumen is on the goat's left side, and some goats will really stuff themselves before

While many parts of a goat are the same as a dog or cat—head, ears, tail, et cetera—there are a few parts of the anatomy that you may not have heard of before.

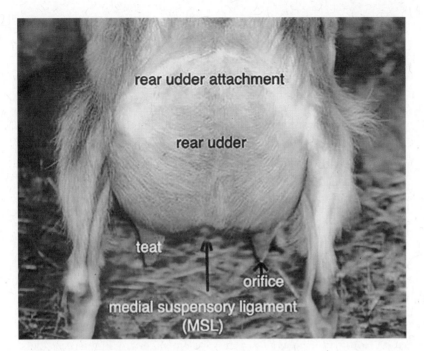

rear udder attachment

rear udder

teat

orifice

medial suspensory ligament (MSL)

A doe with a wide area of attachment in her rear udder, such as this one, will hold up better through the years than one with a very small area of attachment.

they decide to stop eating, which means they can look quite lopsided for an hour or so.

Health

If, like me prior to getting goats, your livestock experience consists of cats and dogs, you are probably accustomed to simply taking an animal to the vet whenever something is amiss. However, if you have more than a few goats, this gets expensive very quickly. And, unlike small animal vets, who can be found in every community, large animal vets are rare and sometimes impossible to find in some areas. For these reasons, many goat owners need to have the information to be able to at least begin to figure out when a goat is sick and whether they can treat it with over-the-counter medication or they need to take it to the vet. Depending on where you live and where the nearest large animal vet is located, a trip to the vet could mean a long drive.

After losing three bucks in a four-month period, I became downright paranoid about my goats dying, and one day I rushed a buck down to the university veterinary hospital two hours away because I thought he looked sick. It was nine o'clock at night, and the on-call vet had been painting her basement when she was called in to the clinic. I don't even remember the buck's symptoms any longer, but I do remember the vet's initial assessment of him as she squatted and watched him walk up to her. She said, "I have a hard time getting too excited about a goat that's walking around, eating, drinking, and chewing his cud." In other words, he would have survived until morning. After lots of tests, he was pronounced healthy, and we went home around midnight. Since then, I always ask myself four questions when I start to worry about a goat's health:

- Is it walking around?
- Is it eating?
- Is it drinking?
- Is it chewing its cud?

If you can answer yes to all four of those questions, the goat is probably fine. "Going off feed" is one of the first signs of many illnesses in goats, and if a goat is unable to stand and walk around, something is seriously wrong.

✧ HUNTER DAVIS, Hawks View Farm Nigerian Dwarf Goats, Harrisonburg, Virginia

This year we experienced our first case of pregnancy toxemia in one of our does. We have a small herd of twelve does who normally graze together, going to and from the barn several times daily to get water, moving about our three acre pasture. At the time, our doe was almost two years old and about to freshen for the second time. She was a month from kidding, approximately 115 days into her pregnancy when we first noticed her with the herd. While the rest of the herd was grazing, she was lying down. I took a mental note and observed. A few more days passed, and I noticed her repeating the lying down behavior—continuing to remain with the herd but lying down among them, resting. It worried me because this was abnormal, especially for her.

After a week of such behavior I called my vet who came over the same day, and diagnosed her with pregnancy toxemia, as I suspected. His way of diagnosis was simple: behavior observation and testing her urine using Ketostix (bought at any local pharmacy). These strips, often used by humans who have diabetes, measure

Condition

Always be aware of a goat's body condition. You should be able to feel the bones, but they should have some meat on them. The only bone that should feel sharp is the withers. Often a good way to assess body condition is to look at the underside of the tail because there is no hair on it, so you can see the skin easily. A goat's tail should look triangular—wide at the base with a good amount of meat on it, tapering into a triangle towards the end of the tail. If there is little or no meat at the base of the tail, the goat is likely underweight. On the other hand, if the tail is meaty all the way to the tip, the goat is probably overweight. It amazes me how fast a goat can lose body condition when it is sick.

Temperature

Taking a goat's temperature is a good first step in determining whether a goat has an infection. Goats do not regulate their body temperature as well as some animals, so most textbooks don't have a single number listed as the normal temperature. Everyone agrees that 102°F to 103°F is fine, but some books say that anything up to 104°F is still normal. In my experience, however, a goat is pretty sick if its temperature is that

---

the level of ketones being spilled into the urine. She measured "heavy" on the color scale on the bottle of test strips, so the vet instructed us to use Nutri-Drench daily (or propylene glycol, which is the main ingredient in Nutri-Drench but is not as palatable) and to continue monitoring her. We did so, giving it to her daily (at a 10 cc dose) depending on the color of the Ketostix. She continually improved and eventually stopped lying around during grazing time. The vet also informed me that does who present with this condition are typically over-conditioned or over-weight, but my doe didn't fit that description. The vet said that in some cases a doe that previously freshened with a single kid and is now carrying multiples can develop toxemia. Thankfully, the month passed and she kidded normally, bearing triplets.

This goat is underweight as you can see from how sunken in her abdomen is. When I ran my hands down her spine, it felt very sharp with no meat on it. It turned out that she had a heavy load of parasites, and within a few days of treatment, she began to regain body condition quickly.

high. A temperature below 101°F also indicates the goat is very sick. A high temperature means the goat's body is fighting something; a low temperature often means the body is shutting down. A low temperature in a doe that is at the end of pregnancy or has recently given birth can also be a symptom of milk fever.

Digestion

New goat owners are sometimes scared by odd noises coming from a goat's abdomen or throat. In most cases you don't hear anything when a goat brings up its cud, but sometimes you do hear a little squeak just before you see the goat start to chew. If you put your ear up to a goat's rumen on the left side of the abdomen, you should hear at least one or two digestive noises per minute, but you will often hear a lot more than that.

Eyelids

The inside of a goat's eyelid should be bright pink or red. If it is pale pink or white, the goat is anemic. Although anemia is most often caused by a parasite infestation, it could have other causes.

Feces

Normal goat poop looks like loose pebbles, beans, or berries. They should not stick together, and the poop should definitely not look like a little log. You should not see anything in the poop that looks like undigested rice or grain, and of course, diarrhea is bad. Abnormal poop can be a sign of parasitism, too much grain, or other problems in the goat's diet.

Grooming

Goats really do not need much in the way of grooming unless you are planning to show. Although some people use dog clippers to trim the hair on the udder, it is a matter of personal preference. As a rule, goats tend to be very clean animals, so they will rarely get dirty enough to need a bath. The only time we bathe our goats is before clipping them because dirt will dull clipper blades quickly. Because we no longer show our goats, we clip them only if one gets lice or in preparation for taking pictures for our website. It is usually impossible to get a good idea of a goat's conformation from looking at a photo of an unclipped animal. Ideally, you want to put your hands on a goat to examine its physical characteristics, but if you are interested in buying a kid from a herd that is a thousand miles away, it is nice to see pictures of the dam and sire after they have been clipped.

Hooves

You should purchase hoof trimmers before you bring home your goats or very soon after. Although someone can hold a kid while you trim its hooves, it is easiest to put an adult goat on a milk stand for its pedicure. Trimming a goat's hoof is a lot like trimming your own nails. Just cut off the extra hoof material, which will usually be curling under each toe. Attempting to cut off enough hoof to create a "perfect" square hoof, which you may have seen in some drawings, can result in profuse bleeding and infection. If you can't make your goat's hooves look perfect, don't worry. Most people don't have perfect feet either.

In general, goats need a pedicure every month or two. When the hooves get overgrown, it is bad for the goat's pasterns, and it makes the job of hoof trimming more difficult. If you go too long between trimmings,

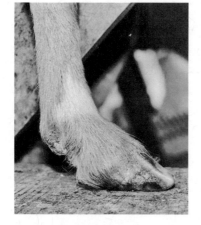

You should be able to draw a straight line down a goat's leg to its hoof. You can see that this hoof is so over-grown that it's forcing the goat's leg into an unhealthy position.

the sole of the hoof starts to grow out, making it difficult to trim off enough of the hoof without cutting into the skin and making it bleed. To correct an overgrown hoof just take off as much as you can without cutting into the soft part under the hoof. As the goat walks around, the sole will get pushed back, and you will be able to cut off more of the hoof in a couple of weeks.

First Aid Supplies

You may already have many of the items you need when it comes to dealing with minor injuries or illnesses with your goats, although you should probably buy additional supplies to keep in the barn so that it is easier to access them when needed.

- **Alcohol:** Even though many people give injections without cleaning the area first, the practice can lead to an infection. You'll be amazed at how much dirt comes off on the cotton swab when you wipe it on the spot you want to inject. You can use alcohol wipes or old-fashioned cotton balls with a bottle of alcohol.

In this picture, a Nigerian Dwarf doe is on a milk stand with her head locked in place. My husband is standing on the goat's left side, lean-ing over her body to hold her body in place, and holding her right rear leg in his left hand. Never point the hoof trimmers in the direction of your hand that is holding the hoof because if the goat kicks at the moment you snip, you could wind up cutting yourself. To be even safer, you can wear a leather glove on the hand that is holding the hoof.

- **Drenching syringe:** A drenching syringe is used for administering oral medications or herbal tinctures. Unlike a syringe that is used for giving injections, a drenching syringe has a long metal tip that is rounded, which makes it easier and safer to administer oral medications and fluids to the back of the goat's mouth, making it more difficult for them to spit it out. A 10 cc syringe is sufficient for most goat medications. It is also a good idea to have a larger drenching syringe available to get water into a dehydrated goat. A 30 cc syringe is roughly 1 ounce, which is a good size for giving water or homemade yogurt, which is rich in probiotics.
- **Gauze pads:** These are 4-inch square pads that are useful for cleaning wounds. Unlike cotton balls, they are sterile. Do not cover an open wound on a goat. Livestock wounds are usually left open to air and drain. Covering a wound with a bandage is a quick route to infection.
- **Hydrogen peroxide:** This is great for flushing wounds. Application is easy if you store this in a spray bottle or use a syringe without the needle to squirt it on an open wound.
- **Syringes and needles:** The 3 cc syringes will work for most goat injections. You will occasionally need a 6 cc syringe or larger. The 20-gauge needles are the most commonly used, 1-inch for IM injections, ½-inch for sub-q.
- **Thermometer:** Of course, you'll want to have a separate thermometer dedicated to use with your goats. Goat temperature is taken rectally. You really should have one available the day you bring home your first goats.

Medicating the Sick Goat

If you have determined a goat needs some type of medication, there are a number of ways to administer it. Each has its advantages and drawbacks.

- **Bolus:** This is the fancy word for giving a capsule or pill to livestock. In spite of their reputation for eating "everything," goats are unlikely to gobble up medication. You can try sticking it down the throat the way some people do with dogs or cats, but your fingers may not come out of the ordeal unscathed. Cattle bolus guns are too large, and dog pill guns made of hard plastic tend to be too brittle and usually break with one chomp between a goat's molars. A dog pill gun made of soft plastic that is somewhat flexible may work. Suppository applicators made for humans also work. Be sure to place the pill at the back of the goat's tongue and hold its mouth closed until it swallows.

- **Drench:** When a liquid medication is to be given orally, instructions will often say to drench. You can use a syringe with the needle removed, or you can buy a drenching syringe (sometimes called a drench gun) that can be used many times and can be taken apart for washing. Medication should be placed at the back of the tongue when drenching so that the goat is less likely to spit it out. Keeping the head horizontal, rather than lifting the nose, is supposed to make aspiration into the lungs less likely.

- **Feed additives:** Medications for coccidiosis prevention and dewormers are available as feed additives. Be sure to read the label because some are fed straight out of the bag while others are mixed with feed. The disadvantage to using feed additives is that you may not know how much of the medication has been consumed if fed in a group, and some goats, especially kids, get so upset about being separated from other goats that they won't eat by themselves. I avoid most feed additives for these reasons and because there are alternatives available as drenches. My only exception is COWP (copper oxide wire particles), which I top-dress on a very small amount of grain for an individual adult. For kids, I bolus COWP.

- **Injections:** Most antibiotics are injected because they tend to disturb the normal flora in the rumen when given orally. Almost all goat injections are given sub-q, or subcutaneously, meaning under the skin, rather than IM, or intramuscularly. Because goats are more likely than some animals to have painful reactions to IM injections, some veterinary professionals say that all injections should be given sub-q. And sub-q injections are considered less dangerous because IM injections in the rear legs can cause lameness if the sciatic nerve is hit when giving the injection. The risk of hitting the sciatic nerve in kids is especially high because the muscle in the leg is so small. Sub-q injections are not without disadvantages, though. There is the possibility of an abscess forming at the site of an injection, which is harmless but may be mistaken for caseous lymphadenitis, often called CL. For that reason, it is a good idea to avoid sub-q injections in the neck, which is the most common site of a CL abscess.

It's important to be aware that drugs are labeled with the recommended method of administration, either sub-q or IM, and that withdrawal times for milk and meat may be different if the recommended method is not used.

A 20-gauge needle works well for most goats, but you may want to use 22-gauge in especially small kids or 18-gauge with medications that are very thick. Whether giving an injection sub-q or IM, always pull back on the plunger after inserting the needle and before injecting the medication to be sure that the needle is not in a vein, which would mean you are about to inject the drug directly into the bloodstream, which could cause death with some drugs. If you get blood when you pull back the plunger, remove the needle and inject it in another spot. It is extremely important to always use a new, sterile needle for each animal to avoid the spread of diseases such as CAE and Johnes, even when you have a herd that has tested negative. After treating with antibiotics, many breeders will give probiotic paste or yogurt with live cultures to re-establish healthy flora in the digestive system. I've never had a goat that was willing to eat yogurt, so I've used a drench gun to administer it orally.

- **Paste:** Supplements such as selenium and probiotics come in pastes. Although there are paste dewormers available for horses, they are not approved for use in goats. Giving a paste is very similar to drenching and bolusing. Place the tip of the dispenser on the back of the goat's tongue to reduce the chance of the animal being able to spit out the paste.
- **Pour-on:** There are pour-on dewormers available, which are simply poured onto the goat's spine, but none are approved for use in goats, although many people use cattle pour-on medications in goats, sometimes as directed and sometimes orally. There have been anecdotal reports of goats dying after being given a pour-on orally.

CHAPTER 5

Feeding Your Goats

✦　✦　✦

Contrary to popular myth, which states that goats will eat everything, including tin cans, goats are actually very picky eaters, and their digestive systems can be upset easily. Goats may have earned a reputation for eating anything and everything because they will eat a lot of vegetation that other livestock won't touch, such as young trees, rose bushes, and weeds.

Fiber-rich foods such as grass, weeds, browse, and hay should make up the majority of the goat's diet. Because goats are technically browsers, rather than grazers, you do not need a piece of property that has pristine pastures. In fact, in recent years, quite a few businesses have been popping up across the United States where goats clear brushy fields that are inaccessible to heavy equipment. Airports in Atlanta, Seattle, San Francisco, and Chicago,[4] as well as the Google headquarters in California[5] are examples of businesses that have rented goats for this purpose. But goats will be happy to eat grass if small trees and bushes are not available. It is important, though, that the grass or browse your goats have access to is green. When drought has turned grasses brown, goats will need some type of additional green feed even if there is an ample quantity of dried grass in the fields. Goats fed brown grass or hay for extended periods of time may wind up with vitamin A deficiency. Bales of hay that are brown on the outside should be green inside when cut open.

Rotational Grazing

Unless you have a hundred acres for four or five goats, consider rotational grazing. It will make your pasture last longer and reduce the incidence of internal parasites in the goats. Rather than fencing in several acres for the goats and letting them spend the whole grazing season on that space, subdivide the pasture into smaller paddocks. The goats stay on one section of pasture for a number of days before being moved to clean pasture with fresh grass, leaving behind their poop, which contains parasite eggs that will hatch and die without a host.

There are a couple of ways you can utilize rotational grazing, and the method you use will depend on the number of goats you have. Four livestock panels can be used to create a 16-foot by 16-foot pen that can be moved every day or two, depending upon how fast the goats eat the grass. This works well for goat keepers who have an acre or two and only a couple of goats. For those with a larger herd and at least five acres, the grazing area is enclosed with a permanent perimeter fence, and temporary electric fencing is used to subdivide the large fenced pasture into smaller paddocks, through which the herd is rotated.

Opinions vary widely on how often the goats need to be rotated to clean pasture, but it depends on weather as well as whether your main objective is pasture utilization or parasite control. For best parasite control, animals should graze an area only once per year, whereas a rotation of every thirty days works if you are only concerned about the best use of the pasture.

The height of the grass also plays a role in deciding when to rotate. Someone once said that goats should never eat below their knees. Technically, they are browsers, not grazers, and they prefer to eat shrubs and young trees rather than grass. Because goats have gone through history not eating off the ground, their parasite resistance is not usually as strong as that of cows and sheep, who do eat off the ground. Goats really should not be eating grass down to the dirt. A commonly encountered recommendation is to move goats to new pasture when the grass is about 6 inches tall. Larvae do not have legs, but they can float up on a blade of grass when it is wet. However, without a true means of movement, parasites don't move very far up on the grass, which is why parasite problems are generally low when goats are consuming grass that is taller. It is a balancing act, though, because when grass gets too tall, it is not as tasty.

Rotational grazing allows you to graze other livestock on a piece of land. Because cows and horses prefer grass and goats prefer bushes and small trees and each species has different parasites, cows or horses can graze the paddock just vacated by the goats. Although sheep and goats have the same parasites, sheep prefer weeds, so they can graze a pasture at the same time as goats. This means you will be able to graze more animals on a piece of land than if you only had one species.

Hay

Dry does, kids, bucks, and wethers usually do well with grass hay, whereas legume hay, such as alfalfa or peanut hay, is commonly fed to does in milk. Finding hay can be a big challenge if you're new to livestock. Although I have heard of a few feed stores that sell hay, in most parts of the country you need to buy it directly from a farmer, and usually you will get a better deal financially if you buy directly from the farmer.

If you can't find hay, there are hay pellets available, although there is some controversy about whether they're as good for ruminants as real hay. For many years conventional wisdom said that ruminants needed the long fibers in real hay, but I've known several people who fed pellets with good results.

Another alternative to traditional hay is haylage or silage, which is partially fermented hay or other plants. Silage has been associated with a number of health problems in goats, such as acidosis, listeriosis, and enterotoxemia. For this reason, it is very important that you do not try to make any type of silage yourself. Without proper knowledge or equipment for making silage, it is possible to wind up with food that could kill goats. It is more complicated than simply raking up cut grass and putting it in a plastic bag. You could wind up growing something toxic.

Commercial haylage is available for purchase, and it tends to have higher protein than dried hay. Although it lasts "forever" in a sealed bag, you need to use it within a week after opening the bag, which is not a problem if you have more than a few goats or if you are also feeding it to other livestock.

Grain

Technically, ruminants should not eat much, if any, grain because it is not the best thing for an animal with four stomachs. However, it can be challenging to keep goats from getting quite skinny in the early months of heavy milk production if they are not fed grain, which is concentrated energy. In fact, you will sometimes see grain referred to as "concentrates." Although grass-fed cow dairies are becoming more common, it is still unusual to find dairy goats raised without feeding any grain.

Jane Wagman of Wags Ranch in Lebanon, Oregon, has a herd of Nigerian Dwarf goats that she has been switching to a grass-based diet. "It does take a while for a goat to develop a rumen for grass-based dairying if they weren't raised that way," she explains. "If they are under a year of age I have had 100 percent success at switching them over. Over a year old my success has been only about 50 percent. Their bodies have stopped growing and just don't adjust as easily. So their production just falls apart because they don't have the ability to support it."

VICKI MCGAUGH, Lonesome Doe Nubians, Cleveland, Texas

I started feeding alfalfa pellets as I hit my 40s. I had always hauled and stacked my own hay, and sold a lot also. I was trying to pitch some heavy bales of alfalfa up to the sixth level when the whole side of the pile fell down, pinning me to the ground in the 90+°F, 90 percent humidity. Instead of picking myself up and doing it again, I sat on a bale and cried. There was no way I would be able to continue doing all this without a better plan. This was in the mid 1990s and I was on the Internet, so I started searching and found a gal in the northwest who did nothing but feed free choice alfalfa pellets from a feeder her husband built for her. She was in a wheelchair, so no way could she lift or feed bales of hay. Her

does were beautiful. I fed out the rest of my hay and started feeding alfalfa pellets instead. The girls balked, of course. They wanted the hay, but with no choice they finally accepted defeat and started eating.

We had also been having problems in the spring with our stored hay. Our heat is oppressive and our humidity is awful, especially in the winters. The feeding quality of our beautiful alfalfa hay was depleted, leaf shattered or molded. The nutrition was waning. By moving to alfalfa pellets, the quality of alfalfa we were feeding was actually better and the girls healthier. The consistency that I was looking for was finally there. Another bonus for feeding alfalfa

There is normally a lot of browse in Jane's goat pasture, but when it dies back during the winter, she supplements with alfalfa pellets. "The ones that have never had a bit of grain in their lives thrive, and their production is above average. Ultrasounds at Oregon State Vet College show extremely healthy and well-developed rumens. The ultrasound pictures of my doe Rita's rumen as a two-year old are now used in classes there to show what a healthy well-developed rumen should look like."

The term "grain" is often used interchangeably with "commercial goat feed." It is uncommon to feed goats only one or two grains, and feeding a mix of grains in the form of a commercial ration is common. However, not all commercial goat feeds are created equal. When we brought home our first goats, we bought the goat feed that was sold at the local feed store, assuming it would be just fine. It wasn't. One of the biggest variables in commercial goat feeds is the amount of copper. It varies from 10 ppm to 80 ppm in the different brands I've checked.

pellets is there is no huge compost pile of wasted alfalfa to clean out anymore. No more money wasted.

My goats are on pasture but mostly on browse. We live on thirteen acres in the Cleveland National Forest. Our goats also have access to several more acres in the buck pen. The girls are eating tiny yaupon leaves and bits of the tender twigs and pine needles. They will eat the pasture grass seed as it blooms out, but just the tender tops. The pasture and browse is also why I don't have to feed very much hay, except when the girls are in the barns for extended periods of time because of rains or hurricanes. I only keep hay out 24/7 when we get our first freeze, usually right before January, until after our last frost in March or April.

The difference in moving from alfalfa hay to alfalfa pellets wasn't dramatic. The girls have very full rumens, great depth of barrel, and milk really well on alfalfa pellets. The difference between feeding grass hay and alfalfa pellets would be much more dramatic. Alfalfa pellets are the perfect protein (our alfalfa pellets are guaranteed 17 percent minimum) and the perfect form of digestible calcium compared with grass hay.

Unfortunately, the feed at our local store was on the lower end, and we wound up with copper-deficient goats.

It is important to read the instructions on the bag of goat feed. Some of the feeds with extremely low levels of copper are meant to be a sole ration for goats because they are little more than hay pellets. Such feeds can wind up being costly because you have to feed far more of it than the more concentrated feeds, which instruct owners to feed a limited amount.

Minerals

When I have questions about how I should be caring for my animals, I often look to nature for the answers. When it comes to nutrition for goats, I've come to realize that few of us in North America can raise goats in a completely natural environment because quite simply we don't have the right environment. Historically, goats lived in mountains and deserts, rather than on grassy plains. They also traveled across many miles. There isn't any evidence that wild goats lived on the prairies of Illinois, which could explain why it has been such a challenge meeting their needs here. Water pumped from deep wells, which have high levels of some minerals, is a factor in a goat's diet. In a completely natural world, goats would not be drinking water that was pumped from a hundred feet below the surface of the ground.

Mineral supplements will make up for the mineral deficiencies that usually occur when domesticated goats live in a place where their wild cousins could never survive. It is important to buy minerals specifically for goats because other livestock have different nutritional needs. Unfortunately, you can still buy "goat and sheep" minerals in spite of the fact that goats will wind up being deficient in copper when consuming minerals that are safe for sheep. Sheep can tolerate about one-fourth as much copper as goats, which means that sheep mineral mixes usually have no copper in them. Even those with copper have an extremely small amount that is not sufficient for goats. Some sheep minerals are also high in molybdenum specifically because it is a copper antagonist and reduces the amount of copper an animal will absorb, which would increase the chances for copper deficiency in a goat that is otherwise getting enough copper in its diet.

Although some people have had good luck with mineral blocks, if you live in an area where your goats need to consume a lot of the minerals, they may have difficulty getting enough from a block because they have soft tongues. Goats also have been known to chip a tooth trying to bite off the minerals on a block. I have used minerals in poured tubs. The minerals in this format are slightly less hard than blocks, but the entire surface was quickly covered with goat teeth marks as the goats were impatient about licking up the minerals and were trying to scrape the surface with their teeth. This is why loose minerals are often recommended as the preferred form of minerals for goats. To provide free choice loose minerals, you need a mineral feeder. If you are not mindful of the height of a goat's back end when deciding where to attach your mineral feeder, you will probably spend a lot of time picking goat berries out of it.

It is important that you never mix minerals with other supplements such as diatomaceous earth, kelp, baking soda, or salt. The minerals in a commercial blend are carefully balanced, and if you add something to them, it can cause a goat's consumption to increase or decrease to levels that could result in deficiency or toxicity of certain minerals. For example, mixing salt or baking soda into a mineral supplement will reduce the goat's consumption of the minerals because you have increased the sodium content. Baking soda can be provided as a rumen buffer, but it should be available free choice in a separate dish so the goats can consume as much or as little as they need.

Most goat mineral mixes already contain 10–40 percent salt, which meets a goat's need for sodium, so you don't need to provide more, especially if you also have baking soda available. Although you can provide separate feeders with additional supplements such as kelp and baking soda, you should not provide a separate source of salt. Salt is used in minerals to control excessive intake. So if you provide a separate source of salt, the goats may not consume enough of the mineral, resulting in deficiencies.

In order to monitor the consumption of minerals, check the label to see how much each goat should consume per day. Multiply the number of goats times the amount recommended times thirty to figure out how much you should be using in a month. For example, one mineral says a goat should consume 0.3–0.5 ounces per day, so a herd of ten goats would consume 3–5 ounces per day, which equals 90–150 ounces or 5.6–9.4 pounds per month. If they are eating less and showing signs of being deficient in some minerals, you can increase intake by top-dressing their grain. On the other hand, if they are consuming more, you can cut back by not keeping the mineral feeders full all the time so that they consume only the maximum amount per month. But if the goats show signs of being deficient after cutting back the amount, you can assume that they really do need to be consuming more than the recommended amount, and you can gradually increase the amount of minerals available until the symptoms of deficiency disappear. Be aware that when goats are first introduced to mineral supplements or have not had minerals for a few weeks, they might consume more than the normal amount for a week or so.

Being faced with a dizzying array of feeds and supplements is confusing. Unfortunately, there is not a one-size-fits-all feed plan for goats because everything from water to hay varies from one farm to another. What works for one farm may not work for another. This is why it's important to watch your goats. They will let you know if your feeding program isn't working. Goats that are well fed will have a shiny coat with soft hair, their body condition will be excellent, fertility and milk production will be high, and kidding problems will be non-existent or rare. If several of your goats have problems with any of these things, odds are good that there is a nutritional deficiency. In a perfect world, pasture and hay are tested and the results sent to a nutritionist, who will custom blend a ration for your goats. Unfortunately, this is usually prohibitively expensive, and there are often minimum amounts of feed that must be mixed by a feed mill for an order, which makes it impractical for those who have only a small number of goats. A simpler solution is to start with a feed plan similar to one followed by someone you know with a herd that is meeting the goals you want to achieve. Then pay attention to your goats and keep records to see if they have the health, fertility, and production that you seek.

CHAPTER 6

Parasites

✧ ✧ ✧

GOATS ARE SUSCEPTIBLE to both internal and external parasites, and parasite control is one of the most confusing subjects for new goat owners because so much of the information that is available is contradictory. Sorting out the difference between chemical and herbal products can get terribly confusing, and the widespread use of drugs and herbs that have never been scientifically tested on goats adds to the confusion for the novice goat owner. This chapter covers the most common parasites that goat owners in North America will face, although there are more.

Internal Parasites

Goats have worms. It is a simple fact that completely shocks and disgusts most new goat owners. Not only do they have worms, but they have other internal parasites as well. The good news is that parasites are host specific, so they don't normally infect people. Goats can and do infect each other, usually by depositing their poop on the pasture, which also happens to be the dinner table, which is why one vet professor said to me, "You will never get to zero parasites with goats."

Although it has been the practice to indiscriminately deworm goats in the past, we have learned that it is smarter to use a dewormer after getting a specific diagnosis. Not every dewormer works to kill every type of worm, so be sure your vet tells you the exact species of worm causing a problem for your goats. Words like "roundworm" and "nematode" are broad categories, not individual species.

Diagnosis of internal parasites requires a microscopic examination of a fecal flotation sample. This is usually done by a vet, but some goat breeders have learned to do fecal analysis themselves so they can keep a close eye on their herd's condition. A positive diagnosis of most worms is confirmed when eggs are found in a fecal sample. Although a fecal egg count can confirm diagnosis, it does not necessarily rule it out.

There are a number of reasons why a goat with a heavy load of worms may have a fecal sample where no eggs are found. Parasites are not evenly distributed throughout the stomach or intestines, and they do not lay eggs constantly, so there will be variability in fecal egg counts from one sample to the next. A common problem with fecal exams is that the sample may not be fresh enough and the eggs may have hatched. Scooping up a few goat berries in the pasture is a waste of time because you have no idea how old they are. Eggs can hatch within an hour or two of dropping on the ground if conditions are right. A fresh sample is best collected by hanging out in the barn for fifteen or twenty minutes after giving the goats hay in the morning. Have a paper cup ready to catch the sample when the goat starts to poop.

Barberpole worm

Haemonchus contortus, commonly called barberpole worm, has proven to be one of the most adaptable internal parasites, and today it is responsible for the deaths of thousands of goats every year. The barberpole worm attaches itself to the inside of the stomach and feeds on the blood of the goat. If present in high numbers in the stomach, these worms will cause anemia. Although the barberpole worm grows to about ¾ inches long, it is almost never visible in a goat's poop. On a necropsy the worms are readily visible in the stomach of an infected goat. Goat extension specialist Steve Hart describes the worms as looking "like coarse hair growing inside the stomach in a freshly dead goat," and notes that the worms will release and can then be seen darting around in the stomach fluid.

Anemia is a symptom of an overload of barberpole worms, although barberpole worms are not the only cause of anemia. The inside of a goat's lower eyelid should be bright red where it meets the eyeball. This is easy to see by pulling down on the skin below the eye until the eyelid rolls out exposing the mucous membrane inside the lid. The lighter the membrane is, the more likely it is the goat is anemic, and if the membrane is white, the goat is dangerously anemic. Similarly, the color of a goat's gums may indicate anemia, but the eyelids are the more reliable place to check as many goats have gums that are always a little pale. If barberpole worm is a problem in your herd, you may want to attend a training session on the FAMACHA program, which provides owners with the latest research in barberpole worms and hands-on training in determining anemia.

Bottle jaw, which is a swelling under a goat's jaw, is often listed as a symptom of a heavy worm load. However, most of the goats I have seen with dangerously high levels of barberpole worms never developed bottle jaw, including some who died. I have seen only a couple of goats develop bottle jaw, and they recovered after treatment. In my experience the absence or presence of this symptom does not necessarily relate to the severity of the parasite infestation.

Goats with a heavy load of barberpole worms sometimes have clumpy poop or berries that stick together, rather than the usual loose pellets. Diarrhea is not usually a symptom of this parasite, and goats with a heavy load can have poop that looks completely normal.

When a doe has a heavy load of barberpole worms, there is often a dramatic decrease in milk production. A goat with a severe infestation will lose weight, ultimately lose its appetite, and eventually die from anemia. A goat that is unable to stand may be so weak from internal blood loss that it is very close to death. This is why it is important to watch for other symptoms and to treat the goat before it becomes severely anemic.

Barberpole worms thrive in a climate of plenty of rain and high summer temperatures. Grass is a necessary element in the life cycle of gastrointestinal nematodes such as barberpole worms. Goats are infected when they consume larvae while grazing in the pasture. The adult worms live in the stomach of the goat, and eggs are shed in the fecal pellets. When temperatures are between 50°F and 95°F, the eggs will hatch into larvae. Although larvae have no means of locomotion, they can float up a few inches on blades of grass wet from rain or the morning dew. When temperatures get too hot, too cold, or too dry, the larvae die.

Brown stomach worm

The brown stomach worm (*Ostertagia* or *Telodorsagia circumcincta*) can infect goats, but it is second to the barberpole worm in terms of prevalence and mortality. It feeds mostly on the nutrients in a goat's stomach, and it can damage the lining of the stomach, causing poor digestion. Unlike the barberpole worm, it does not cause anemia. A goat with an infestation of brown stomach worm can have a poor appetite, grow slowly, get diarrhea, and sometimes bottle jaw. This worm also prefers wet conditions, but can tolerate cooler temperatures than the barberpole worm. It is more likely to be a problem in spring or fall rather than during the hot summer months. Goats are infected with the brown stomach worm while grazing in pastures where other goats have left infected fecal pellets.

Bankrupt worm

The bankrupt worm or black scour worm (*Trichostrongylus colubriformis*) is a worm that lives in the small intestines. The eggs are shed in the fecal pellets. Larvae prefer outside temperatures around 70°F to 80°F. Goats become infected when they eat grass from a contaminated pasture. It can cause diarrhea, but rarely kills goats. It interferes with a goat's ability to get adequate nutrition from food, however, causing a goat to have very poor body condition, appetite, and production.

Tapeworm

Moniezia, or tapeworm, is one of the few worms that can be seen with the naked eye. Poop that looks like it has rice or noodles in it usually indicates an infestation of tapeworm. This worm is the most upsetting to owners because it looks so disgusting. As one vet said to me, they are far worse for the mental health of the owner than they are for the physical health of the goat. Tapeworms usually do not cause a problem in an adult goat unless the infestation becomes so heavy that it causes an intestinal blockage. Growth of kids infected with tapeworm can be slowed somewhat because tapeworms absorb nutrients from the food in the goat's intestines, but I have seen very thrifty, well-conditioned kids pass tapeworms.

These worms infect goats through an intermediate host. Goats deposit the eggs on pasture, as they do with other worms. But in this case, field mites or grass mites, which are similar to chiggers, consume the eggs. The mites live on grass and browse, and goats become infected by grazing. Because the worms live in an intermediate host rather than simply on blades of grass, a pasture will usually remain infected with tapeworms from one year to the next, which is much longer than other worms.

Liver fluke

Liver fluke (*Fasciola hepatica*) is another parasite that needs an intermediate host to infect goats. In this case it is a snail, which the goat ingests while grazing. The fluke then migrates to the goat's liver. The main symptom is anemia with weight loss. Sometimes it causes death. This parasite is usually a problem in wet areas and is why some people don't let their goats graze around creeks and ponds, although snails can be found in other places. Snails are most active in spring, however, so ingestion is less likely during other times of the year. It takes a couple of months for the fluke to mature in the liver and begin to cause symptoms.

If you have deer and snails in your area, you may also have a problem with *Fascioloides magna*, another type of liver fluke, which is carried by deer, and can be fatal to goats. Deer flukes are not a common problem in goats, but discouraging deer from entering the pasture helps prevent infection. A livestock guardian dog will keep deer and predators away.

Lungworm

There are several types of lungworms (*Muellerius capillaries, Protostrongylus rufescens, Dictyocaulus filaria*), which infect goats when they ingest the larvae. Rather than staying in the digestive tract, lungworms migrate to (you guessed it) the lungs. You will see first-stage larvae, rather than eggs, in a fecal sample. The main symptom is a chronic cough, but like other parasites, lungworm also takes a toll on the goat's body and causes poor production and weight loss. It sometimes causes death. Some types of lungworms require a snail as an intermediate host, though not all do. Treating lungworm may require two doses of dewormer given thirty-five days apart.

Meningeal worm

Although it is more common in llamas and alpacas, meningeal worm (*Parelaphostrongylus tenuis*) can infect goats. It is a common parasite in deer and uses the snail, ingested by goats when grazing, as an intermediate host. In goats this worm migrates up the spinal column to the brain, causing paralysis and eventually death. As with liver flukes, goats usually ingest the snail in the spring, but it takes three or four months for the meningeal worm to migrate to the brain. Symptoms are generally neurologic and include leg weakness, paralysis, circling, holding the head to one side, tail twitching, sitting like a dog, and blindness. There is no definitive diagnostic procedure for determining if a goat has meningeal worm, so it is usually determined by simply looking at symptoms, which are similar to listeriosis. Although a spinal tap works well for diagnosing meningeal worm in llamas, it does not work well with goats. Years ago I had a goat with these symptoms, and the vet simultaneously treated for listeriosis and meningeal worm. The goat made a fully recovery.

To prevent problems, discourage deer from coming into your pastures and do not let goats graze in swampy areas. A livestock guardian dog usually will keep deer away, in addition to predators.

Coccidia

Although usually lumped into any discussion of worms, coccidia are protozoa that infect the small intestine. Adult animals have immunity to coccidia acquired from exposure early in life and so are not bothered by the presence of small numbers of coccidia in the intestine. Coccidiosis, which is the illness caused by coccidia, is most often a problem in kids over three weeks of age, and is the most common cause of diarrhea in kids, although not the only cause. A severe case of coccidiosis in a kid can result in permanent scarring of the intestine, which will inhibit absorption of nutrients and cause slow growth. If not treated, kids can die from dehydration caused by diarrhea. If you do not see an improvement within a couple of days after starting treatment for coccidiosis, talk to your vet about the possibility of cryptosporidiosis.

The protozoa can reproduce asexually in the intestines, so once a kid is infected and symptomatic, it needs to be treated. Because coccidia are often lumped into discussions with worms, some people are under the mistaken impression that dewormers will treat coccidiosis, but they do not. A five-day treatment of a sulfa drug is commonly used to treat coccidiosis in goats. Some people will use a lower dosage in the goats' water for two weeks or longer as a preventative. This is a bad idea because if goats are consuming a small amount of this drug and still become infected with coccidiosis, the sulfa drugs will not work to kill the coccidia. Just as with antibiotic resistance and dewormer resistance, coccidia can become resistant to drugs with continual exposure.

Amprolium is approved for the treatment of coccidiosis in cattle and poultry. It is not labeled or approved for use in goats, but it is used as a coccidiostat for goats. It is not recommended for use as a preventative because of the risk of long-term use causing a thiamine deficiency. Medicated milk replacer, minerals, and feed containing decoquinate or monensin are available as preventatives in goats. Oregano oil looks promising for treating poultry with coccidiosis, and there are anecdotal reports of black walnut hull powder being used in goats, but more research is needed before we can definitively say if these remedies work consistently in goats and at what dosage. Good barn hygiene, however, should be the first line of defense against coccidia.

Controlling Internal Parasites

It takes a comprehensive approach to control parasites in goats. You cannot do it by simply using dewormers, either chemical or herbal. The goal is not the elimination of worms in a goat's body. Because zero worm load is impossible, it is important that goats develop some natural resistance to worms, and that does not happen unless there are some worms in the body. The key, however, is not to have so many worms in a goat's body that the worm load overwhelms it, affects growth or production, and ultimately kills the goat. It is more effective to be proactive with your management, rather than to be reactive, trying to save ill goats with drugs.

Many new goat keepers have no problem with worms in the first couple of years because the pasture is sparsely stocked. They don't realize that the low stocking density is the reason for the absence of problems and assume that whatever they are doing in terms of deworming (or not) is working. This explains why a deworming protocol such as pour-on chemical dewormers and some herbal preparations appears to work even though research has shown that it does not work. If there is no worm problem to begin with, "nothing" may have worked just as well. As the herd grows, however, the parasite load on the pasture increases. A dead goat is usually the first indication of a problem when a goat owner believes parasites are not a problem on their farm. It happened to me, I've seen it happen to other people on my Internet goat forum, and it is often mentioned in scientific journal articles about goats and worms.

Dewormers

Although it appears there are about a dozen chemical dewormers on the market, they all fall into one of three categories, and drugs in the same category have a similar mode of action and work to kill similar worms. As of this writing, there is a new dewormer that will be the first in a new class that has been submitted to the US Food and Drug Administration (FDA) for approval, but monepantel is not expected to be on the market in the United States for a couple of years. It is already available in Australia, though.

"White dewormers," so called because they are most commonly sold as a white liquid or paste, are benzimidazole dewormers, which include fenbendazole and albendazole. It appears there are more dewormers than there really are because some brand name drugs have the same one or two active ingredients. Panacur and Safe-Guard both use fenbendazole as the active ingredient. Valbazen is the trade name for albendazole. Only Safe-Guard is approved for use in goats, but other drugs are used off-label. The benzimidazole class of dewormer is the only one effective against tapeworms.

The second class is the macrocyclic lactone dewormers, which are sometimes called the clear dewormers because they are a clear liquid or gel. This includes ivermectin and moxidectin. Moxidectin is considered much stronger than ivermectin, so ivermectin is generally used first

Extra-Label Drug Use in Goats

Because goats are a minor species in the United States, it is not profitable for drug companies to test their drugs on goats to get approval from the Food and Drug Administration (FDA) to have the drugs labeled for use in goats. Veterinarians and owners, however, need to be able to treat a sick animal. The Animal Medicinal Drug Use Clarification Act (AMDUCA) of 1994 created guidelines for using drugs extra-label so that drugs can be used legally in different species, even when the drug has not been FDA approved for that species. However, this does not mean that you can use whatever drug you want to use in goats.

According to AMDUCA, you can use an extra-label drug to treat your animal only after consulting with a licensed veterinarian and only if there are no FDA-approved drugs available to treat that condition. You may also use an extra-label drug when the only approved drugs don't work. For example, this means that legally you should only use fenbendazole or morantel tartrate as dewormers because they are labeled for use in goats. However, if they no longer work, you could use a dewormer that is labeled for a different species.

To find approved dosages, as well as meat and milk withdrawal times for extra-label drugs, consult the Food Animal Residue Avoidance Databank (FARAD), www.farad.org, which is supported by the United States Department of Agriculture (USDA) and maintained by several universities. If you don't see a milk or meat withdrawal time, you should not use that drug in an animal that will be used for milk or meat. Sometimes there is no recommendation because the drug has not been studied, but it may also be because the drug stays in the animal's system for so long that it is impractical to use it, especially in dairy animals.

because moxidectin can still be useful in some goats when ivermectin no longer works. Although these drugs are not labeled for use in goats, they are approved for extra-label use orally or as a pour-on. Moxidectin injectable is not supposed to be used in goats, and although ivermectin injectable is approved for extra-label use, the milk withdrawal is 40 days, making it impractical for use in milkers.

Imidazothiazole dewormers are the third class of dewormer. They are usually sold in a solid state, such as a water-soluble powder, a bolus, a medicated feed, or a feed additive. This class includes levamisole and morantel tartrate. Morantel tartrate is the only drug in this class approved for use in goats. Levamisole is sold as a bolus or as a powder that can be mixed with water and given orally. Because levamisole has been used so rarely in goats, some people have found that it works when other dewormers are no longer effective, including Rumatel, which is in the same class. One must be very careful, however, when using levamisole because the margin of safety when dosing is very small, making it easy to overdose a goat. While most sheep dewormers are used in goats at twice the sheep dosage, levamisole is used at only 1.5 times the sheep dosage.[6] Signs of toxicity include tearing of the eyes, excessive salivation, and the goat walking like it's drunk. In most cases, the goat will recover after a few hours or days of rest.

Dewormer resistance

It is a well-known fact that internal parasites are the leading cause of death among goats. Unfortunately, it is a problem of our own making. You have probably heard of antibiotic-resistant superbugs, which resulted from the overuse of antibiotics. A similar phenomenon has occurred with chemical dewormers and internal parasites in livestock. Veterinary professionals thought parasites could be eliminated in livestock and began recommending routine use of chemical dewormers in healthy animals. Although this worked in the short term, the long-term result has been dewormer-resistant parasites.

During the 1980s and 1990s, the standard advice was to deworm livestock on a pre-determined schedule in an attempt to keep internal parasite loads so low that they would never adversely affect production.

When people started to see that the parasites were surviving dewormings, the advice was changed slightly to tell people to switch dewormers from one treatment to the next so that the new drug would kill the worms that weren't killed by the previous dewormer. To reinforce this idea one company even came out with a horse dewormer system that numbered their different dewormers so owners could use them based on which quarterly deworming they were administering. To complicate matters for the goat owner, large animal vets see plenty of horses and may focus their continuing education on the animals that make up the largest portion of their practice. This means that when it comes to goats, many vets are still repeating what they learned in vet school fifteen or twenty years earlier. Depending on where a vet went to school and who taught the small-ruminant classes, even some newer vets are still practicing with old information. One researcher told me that about half of what we know about parasites has been discovered in the last twenty years and much of that was not taught in vet school until the last five years. Many of the old recommendations were based on personal experience, opinions, and logic and subsequently have been proven to be wrong.

In the last ten years, a lot of research has been done on internal parasites, especially on barberpole worms, because of the large losses experienced by sheep and goat producers in the southeastern United States. A great deal of research in this area, however, has been done in Australia, New Zealand, and South Africa. One study after another clearly shows that dewormer resistance is a real problem. Because there are only three categories of dewormers, it does not take long for a resistance problem to develop when they are used frequently. Research shows that if parasites are resistant to one drug in a class, the other drugs in that class may not work.[7]

In addition to being used for internal parasites, ivermectin and moxydectin are sometimes used as a pour-on for external parasites, such as lice and mites. Unfortunately, the internal parasites are being exposed to the dewormer when it is used as pour-on, and it only kills about 50 percent of the internal parasites when used in that manner, which means it leads to problems with resistance faster.

When talking about resistance, it is important to understand that we are not usually saying that the drug kills zero percent. Ideally, a dewormer will kill as close to 100 percent as possible. But the more you use a dewormer, the fewer worms it will kill because the population of resistant worms is growing and reproducing. Depending on how sick a goat has become from a heavy load of parasites, a reduction of 70 or 80 percent may not be enough to save it from dying. It is also possible that a severely debilitated goat will die even if all of the worms are killed. The sooner you stop over-using chemical dewormers when parasite resistance is evident, the better your chances of reversing the problem will be.

Common deworming practices

Using dewormers selectively means that you deworm only when an animal has a level of worms that is causing it to be anemic, not grow normally, not produce milk as expected, or otherwise show signs of illness. You do not deworm a doe simply because she has kidded or because other goats in the herd have a high level of parasites. These two practices will lead to rapid dewormer resistance problems. Now, you're probably wondering why so many people do prophylactic deworming and why many vets recommend it. Without looking at long-term research, prophylactic deworming seems to work.

Common Practice #1: *If you deworm all the does after kidding, they have lower parasite loads, lose less weight, and produce more milk.* Why wouldn't everyone want to do this? Because in the long term, this is part of a losing strategy. It is unlikely that all of the does needed deworming, and using a dewormer when it isn't needed means you have just taken one more step towards dewormer resistance and the day when a dewormer will be needed and it no longer works. By deworming all of the does, you are selectively breeding worms for dewormer resistance. Also, by deworming routinely at kidding, you lose the opportunity to select breeding stock based on parasite resistance.

Common Practice #2: *If you have a couple of goats with a high parasite load and you deworm the whole herd, no other goats wind up with a high parasite load.* Why wouldn't everyone want to do this? First of all, the rest of the herd probably would not have gotten sick, even if you had not given them a dewormer. I've heard vets say that it's only common sense that if a couple of goats in a herd have high worm loads, the rest of them do too. Unfortunately, common sense fails this time. Research has shown that in most cases only a few animals in the herd will have a high parasite load. Eliminating nearly all worms works well in the short term, but if you want to have goats long term, you will eventually face the consequence of dewormer resistance.

If you deworm the whole herd, you kill every worm that is susceptible to that particular dewormer. Sounds great at first. Unfortunately, no dewormer kills 100 percent of worms. This means that every worm left living in your goats is resistant to the dewormer that you just gave them. The only reason a dewormer is effective more than one time when you do a whole-herd deworming is that the goats are continuing to ingest worms in the pasture and those worms are not resistant. But they mate with the resistant worms inside the goats, and some of those offspring are resistant to the dewormer. So, the more you use a dewormer, the greater the number of resistant worms you are breeding.

Common Practice #3: *Deworm your goats ten days after the initial deworming to kill parasites that have just hatched because the first deworming will not have killed the eggs.* Of course, when you do that, you do kill some worms that have hatched, but you also wind up with more resistant parasites that survived the second deworming, increasing the number of resistant parasites on your pasture. The more often you use a dewormer, the faster you are breeding the parasites to be resistant to the dewormer you are using. This advice for a follow-up deworming was originally given for fenbendazole, which had poor efficacy against arrested worms, which would then be killed by the second dose a couple weeks later. Unfortunately, a lot of people concluded it would be a good idea with all dewormers, regardless of whether it was necessary. After using any dewormer, you should continue to monitor the body condition and anemia status of the goat. If the animal does not improve, follow up by doing a fecal exam. If the reduction in fecal egg count is minimal, you may need to give a second dose of dewormer or to use a different dewormer.

What I learned from Tennessee Williams

During chore time in the middle of a freezing winter, I saw one of my bucks fall down when another goat walked past him and barely touched him. That was obviously not good. I separated him from the other bucks so they wouldn't bother him and so that he wouldn't give them whatever was making him sick. I checked his eyelids, and they were white, so I did a fecal analysis. The slide was covered with eggs from the barberpole worm. I gave him a double dose of the herbal dewormer I had recently purchased online in my quest to find something that would work since the chemical dewormers had stopped working a couple years earlier. After several days of giving him the herbal dewormer and seeing no outward improvement, I did another fecal, and the slide was again covered with eggs.

In my head, the voices of vets from the past were telling me to give all the bucks a dewormer *because it just makes sense that if this one is so badly infected, the other ones are too*. But I had been reading a lot of research, and instead of deworming, I went into the pen and checked eyelids for symptoms of anemia, and then I waited for them to poop. I collected a fresh sample and did a fecal on a buck, as well as two does. I found very few parasite eggs on any of the slides. I was moving the slide from side to side and finding a solitary egg here and there for a total of two or three eggs on each slide. Thinking that this was an unlikely result, I repeated the exams on fresh poop and got the same results. Seeing that none of the other bucks were anemic and finding a negligible number of eggs in the fecals, I hesitantly decided not to give the other bucks any type of dewormer.

Unfortunately, Tennessee continued to decline over the next week. He could no longer stand at all, and I needed to move him to clean straw regularly because he was peeing and pooping where he was lying. I gave him a chemical dewormer, but it didn't reduce the worm load either. I was amazed by his ability to hang on because most goats are either on the mend or dead within a day or two of going down. Maybe the herbs had boosted his immune system, even though they didn't actually kill the worms. Maybe he was just stronger than most because his dam is the most parasite-resistant goat in the herd. In the end, the worms prevailed, and Tennessee died. As the weeks passed, the physical condition of the other bucks remained robust, and none of them required a dewormer.

The fewer the goats that you use a dewormer on, the better the chances are of the herd never developing dewormer resistance because the proportion of the resistant parasites will be small compared to the total population of non-resistant worms on the pasture and in the goats. For example, if you have a herd as small as ten goats and you deworm only one goat, killing 90 percent of its worms, and this goat had 25 percent of the worms in the herd, only 2.5 percent of the worms (10 percent of 25 percent) in the herd would be resistant to the dewormer you used. Therefore, only 2.5 percent of the worms will be producing dewormer-resistant babies, compared with 100 percent of the surviving worms in every goat if you had dewormed the whole herd.

If you have already experienced dewormer resistance within your herd, there is hope. Several years ago researchers were saying that once a herd had a resistance problem, it would last forever. However, in my experience, it can be reversed. After having three goats die from parasites in four months, I began to use chemical dewormers faithfully, and I rotated them regularly, which was the standard advice at that time. Within about four years, nothing worked well, and my goats were dying in spite of using chemical dewormers. At that point I got very serious about doing whatever I needed to do to avoid having a parasite overload in the goats, which included rotating pastures and kidding during the dead of winter when parasites are mostly dormant. My use of dewormers plummeted. After about three years, I had a goat with a heavy load of parasites, and when I used a chemical dewormer, it caused a significant reduction in the fecal egg count.

Alternative Dewormers

Because of the increased interest in organic agriculture, as well as the problems being experienced with dewormer resistance, there is quite a bit of research being done on a variety of alternative dewormers.

COWP

One of the most promising natural dewormers is copper oxide wire particles (COWP), which are sometimes referred to as copper boluses. They are sold as capsules filled with extremely tiny bits of smooth cop-

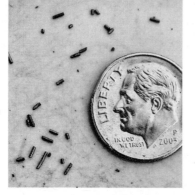

Copper oxide wire particles sound scary to some people, but you can see in this photo that the particles are extremely tiny. They are not sharp, and they dissolve in the goat's stomach over the course of a month.

per. Once the copper wires are in a goat's stomach, they slowly dissolve over a few weeks. More than a dozen studies have been done over the past few years, and the results are fairly consistent, showing that using COWP significantly reduces stomach worms (such as barberpole).[8] It has not been shown to be effective against intestinal worms.

I use COWP in adult goats that are borderline anemic, which is a symptom of barberpole worm. If a goat is severely anemic, however, I also use a chemical dewormer. I give COWP to all kids that are weaned, and although a few kids wind up needing a chemical dewormer, most of them don't. Because doelings are never weaned if we are keeping them, I give them a dose of COWP at around six months unless I see signs of anemia or copper deficiency earlier than that. The stress of moving can often cause an overload of parasites in kids, so I discuss this with buyers and ask if they want me to give the kid a chemical dewormer prior to or at the time they pick up the kid.

Some older guides to natural care of livestock recommend copper sulfate for animals that have a high parasite load. Although some people have reported success with this, others have had animals die from copper toxicity. Copper sulfate is extremely well absorbed, whereas copper oxide is not. COWP sit in the stomach and gradually dissolve over the course of a month. The case studies I've seen in the literature about copper toxicity have all been in goats that were receiving copper sulfate supplements. I've read more than a dozen studies done with COWP being used for worm control, and none of the goats ever died from toxicity. In fact, in some of the studies, researchers checked liver levels of copper in the goats after treatment, and none had dangerously high levels of copper.

Herbal dewormers

The most definitive thing we can say about herbal dewormers is that we don't know exactly what works yet. Although they appear to work for some breeders, they don't work on every farm, and sometimes they don't work on different goats on the same farm. To say that something "works," it needs to work predictably. I use many herbs myself, and I know they work based on personal use and scientific research. Ginger reduces nausea, and senna is a laxative. Willow bark is a pain reducer, and is, in fact, the original source of the active ingredient in aspirin. Scientific study has focused far more on the effects of herbs on humans than on livestock. Although you can often use the same herbs for the same purpose in animals as in humans, there is not much research on using herbs for internal parasites in people, and there is no research on the parasites that are specific to goats. I've read as many studies as I can find on herbs, and I've tried a variety of different herbal treatments for parasites without finding anything that works predictably.

The main anthelmintic herbs are in the genus *Artemisia*, and most researchers have not been willing to study them as a dewormer because wormwood (*Artemisia absinthium*) has a bad reputation. It can be addictive and hallucinogenic in humans, and long-term use can cause liver damage. Historically it may have been used to induce abortions, so most herbalists say it is not safe to use during pregnancy. Wormwood is usually one of the ingredients in herbal dewormer preparations that you can buy online, but the level of wormwood in herbal products may be too small to make a significant difference in a goat that has a dangerous level of internal parasites. The little research done with wormwood has shown that a goat needs to ingest a fairly large amount of the herb to reduce the parasite load. Mugwort (*Artemisia vulgaris*) is closely related to wormwood, and some people say it has the same anthelmintic properties without the negative side effects, but research is needed to verify these anecdotal reports.

Although there are several herbal combinations sold by individuals online, none of them have scientific studies to prove their effectiveness. When researchers did conduct a scientific study where they treated half of the goats with a herbal dewormer and half with nothing, they found no difference between the two groups.[9] The lead researcher said that the amount of herbs used was probably too small to be effective, which is possible because the directions called for only a tablespoon of the herbs to be fed to a goat. Most studies that do show some effect on worm load have much higher levels of herbs, including one study that used one pound of wormwood per goat. Using that amount of wormwood, however, worried most researchers because of wormwood's reputation for damaging the liver. And unless you are growing the wormwood yourself, that would be very expensive even if only a few goats needed it.

One challenge in working with herbs is that the strength of herbs may vary from one batch to the next, which may explain why I have had different results using wormwood at different times. Growing conditions, storage, and age of an herb can affect its potency. Some herbal companies sell "standardized" herbs, but there is some disagreement about whether that solves the problem or creates a new one. A standardized herb has been tested and manipulated so that "active" constituents are at a guaranteed level. However, in many cases, we don't know what the active constituent is, and even if we think we do, it is possible that other constituents are also important. By altering one component, we may be inadvertently causing an imbalance in the way that the herb works.

Diatomaceous earth

Diatomaceous earth, often called DE, is a natural insecticide that has proven to work well in the garden, and people have argued for years about whether it is effective as a dewormer for internal parasites in livestock. DE is the fossilized remains of marine diatoms, and it works in the garden by dehydrating insects because it is very absorbent. It also appears to be very abrasive because when looking at it under a microscope, you can see that it has very sharp edges, even though it feels like talcum powder or cornstarch. The assumption that the sharp edges cut parasites and kill them is incomplete. DE works by absorbing several times its weight in liquid, so when it comes in contact with garden insects or even your skin, it draws out moisture. This is why DE kills bugs in the garden that would appear to be immune due to their protective exoskeleton. Although DE's abrasive properties don't damage the exoskeleton, its absorptive properties will absorb the waxy coating on the exoskeleton, which can kill the bug.

Because the inside of an animal is a wet environment, DE will absorb as much liquid as it is capable of holding from the goat's digestive system before it ever comes in contact with a parasite. So DE cannot kill internal parasites by dehydration. Can DE damage internal parasites by essentially stabbing them to death with its sharp edges? We don't know for sure. Some people are under the mistaken impression that DE gets soft when it gets wet. This is not the case, however, as DE has been used in toothpastes and in facial scrubs because of its abrasive characteristics even when mixed with liquids.

If DE is going to work, it has to be given in large quantities because it is a physical insecticide, meaning that it has to come in contact with a parasite to kill it, so putting a teaspoon of DE on a goat's grain would not be nearly enough. Those who swear by DE's effectiveness say they add ¼ cup or more to their goat's grain, twice a day for several days. Studies on the effectiveness of DE have had mixed results. One study concluded that DE in the diet dried out goat pellets faster, effectively killing the larvae, which decreased risk of reinfestation. However, when the same researcher repeated that study two summers in a row, the results were not consistent.

There is a lot of anecdotal evidence on both sides of the DE argument, and the recommended dosage varies from one source to another. The first time I used DE, I simply added a little to my goats' minerals, which is a fairly common recommendation on the Internet. However, that means individual goats are getting an amount of DE that is so small as to be almost immeasurable and completely ineffective. In fact, two goats died from parasites while I was doing that. If you are going to try DE, be sure to buy food grade and not the variety that is used in swimming pool filters.

I am not completely convinced that DE works, but one thing I have noticed in my own use of DE is that when animals have a high parasite load, they love to eat it. Its chemical composition is mostly silica, so perhaps those goats have a high need for that substance. Or perhaps because anemic goats are essentially starving, they want to eat everything. There is still a lot of research that needs to be done in this area.

In the meantime, all of us should be taking detailed notes when we try alternative dewormers so we can figure out what works with our herds and what doesn't.

Preventing Infection

If fighting internal parasites with dewormers is ultimately a losing battle, what can you do to keep your goats healthy and productive? There are a number of environmental controls that can be utilized to eliminate parasite problems almost entirely. If goats needed dewormers to survive, they would have all become extinct before dewormers were invented. Many of the following ideas are not new, simply rediscovered from the era before drugs were developed. There was an old Scottish shepherd's saying that one should "never let the church bell strike thrice on the same pasture,"[10] meaning that sheep should be moved to new pasture at least every couple of weeks.

Pasture rotation

Pasture rotation is one of the most important management tools you have in preventing parasite problems. Worms and coccidia can contaminate a pasture and cause continual reinfestation of goats eating the grass. The shorter the time goats spend on pasture before being moved to fresh ground and the longer they stay off previously grazed pasture, the fewer parasite problems you will have. For example, if you could give goats a clean pasture every day and never put them back on any section for a full year, you would probably never have any parasite problems. Of course, few people have that much pasture or the time to rotate their goats every day, especially if they have a large herd, which may require the help of several people to move. Ideally, goats should be moved off a piece of grass within five to seven days and should not return to that spot for six weeks. However, the more often you can move them and the longer they can stay off of a piece of ground, the better.

Weather

Weather plays a big role in the rotation schedule. If temperatures are above 95°F and it is very dry, you will be able to leave goats on a piece of pasture much longer than if it is 70°F and raining every other day. You can usually let goats graze in a pasture until you get at least ½ inch of rain, which provides the perfect environment for larvae to thrive. When temperatures are in the upper 90s, you can probably let a pasture rest for as little as five or six weeks. However, keeping goats off pasture for at least three months in more moderate temperatures is necessary to help prevent ingestion of worm larvae left on the pasture during the previous grazing period.

Grass height

Height of grass is an important consideration when goats are on pasture. Because most larvae are on the lower 2 or 3 inches of grass, it is a good idea to move goats to the next pasture when the grass is about 5 or 6 inches tall in their current pasture and definitely by the time it is grazed down to 4 inches. Early in the grazing season, some varieties of grass may not get very tall, so you will have to move goats to fresh pasture sooner than you will have to do later in the grazing season.

This is one reason it is better to create many smaller paddocks and move goats more frequently. Goats in a very large pasture will completely ignore some areas and overgraze others, and the longer the goats are there, the worse the problem gets. Young grass is tender and sweet; taller grass begins to grow tough and becomes less appetizing. Sections of untouched grass can be mistaken for an abundance of grass in the pasture, but unfortunately, goats will keep going back to the grass that is extremely short and covered with infective larvae.

Mixing species

Integrating other species such as poultry, horses, cattle, or pigs into the pasture rotation can help you utilize pasture more effectively than if you only have one species. Grass is at its most nutritious about thirty days after it was last grazed or cut, but that is also when larvae tend to be at the most highly infective stage. However, horses, cattle, and pigs are not susceptible to worms that infect goats and are able to digest the larvae. In other words, cattle, horses, and pigs can clean up a pasture, making it safe for goats to graze sooner than would otherwise be safe.

Using sheep, alpacas, or llamas is not effective, however, because some goat worms can also infect them.

Browse

Providing goats with areas to browse is also an effective tool for controlling parasites. Goats have a much harder time dealing with parasites than sheep and cattle because as mentioned earlier, goats have not been grazers throughout history, like sheep and cattle. Although goats will eat grass, they are browsers and prefer to eat small trees and shrubs, which have no parasite larvae on them.

Dry lot

It might be tempting to keep goats on a dry lot or in a barn to avoid grass and the inevitable parasites on it. However, when goats are kept inside, the bedding needs to be kept cleaner than you would expect, in order to avoid coccidiosis in kids. If a stall is too dirty for you to sit down in, it isn't clean enough for kids. A dry lot needs to be large enough that there isn't a build-up of manure in it. Even in a dry lot a goat can become infected when it sticks its head through the fence to eat the grass that has been infected by other goats pooping along the fence line. Also keep in mind that not all parasites need grass. Intestinal thread worm (*Strongyloides papillosus*) can hatch in bedding or soil and infect a goat through its skin, although it is not a common parasite.

Breeding for resistance

Selecting goats for parasite resistance is one of the most sustainable ways to deal with parasites. Some goats are clearly more parasite resistant than others, and research is under way to see if some breeds are more resistant than others. In the meantime, some researchers and veterinarians are encouraging breeders to cull animals that have an unusually hard time with parasites. By eliminating goats from your herd that need multiple dewormings, you reduce your dependency on dewormers. I know one vet who raises goats and will cull an animal the second time it needs a dewormer. In my own herd, I breed goats with mediocre parasite resistance to goats with high parasite resistance.

Seasonal birthing

It has been a common practice for many years for goat breeders to give a chemical dewormer to every doe as soon as she kids. An increase in worm load is thought to be a normal part of freshening, and the dose of dewormer at that time has been considered unavoidable. However, some breeders are starting to question this practice and are waiting to deworm until they see that a doe truly needs it. There is no doubt that some does have an increase in parasite load after kidding and they do require a chemical dewormer. There are alternatives for some goats, though.

Most of my goats do not require dewormer after kidding if they give birth in January or February when parasites on the pasture are frozen or in late summer or early fall. If they give birth between March and June, however, the odds are not in their favor. July is variable based upon whether it is a dry or rainy summer. A number of breeders and researchers have begun to make similar observations about avoiding dewormers by timing kidding so that does are not freshening when parasites are at their highest level on the pasture. Keep in mind that the ideal time for kidding can vary from one location to another, depending upon your weather.

Tannin plants

In addition to traditional herbs, there are a variety of tannin-producing plants such as sericea lespedeza, birdsfoot trefoil, and chicory that have been shown to reduce parasite loads. These are plants that grow wild in many places but can also be planted in your pasture. However, it is not as simple as seeing a few of these plants growing in the pasture and assuming that you don't have to worry about parasites. Sericea lespedeza is the most studied of the natural antihelmintics, and it has been shown to reduce parasite levels at varying degrees. Goats grazing on fields of lespedeza daily have been shown to have no parasite problems. However, as their grazing time decreases, the parasite problems increase. There are currently no solid recommendations for how much lespedeza goats need to consume to control parasites.

As you have probably realized by now, there is no magic bullet. Controlling parasites has to be a multipronged approach. And just when you think you have it figured out, things can change. This is why it is important that you understand the life cycle of the parasites, rather than simply memorizing a particular deworming protocol. An unusually mild winter or wet spring can make problems worse by providing the perfect environment for larvae to survive on pasture for an extended period of time. A drought may improve the problem by drying up the larvae on pasture. Raising goats sustainably doesn't mean substituting a natural dewormer for a chemical one. It means creating a management plan and an environment in which goats can thrive with the fewest external inputs from humans.

External Parasites

A variety of lice, mites, ticks, and fleas can affect goats. Some are simply a nuisance, but others can cause health problems.

Lice

Chewing lice crawl around through the goat's hair eating dead skin and other debris, whereas sucking lice attach themselves to the goat and feed on the goat's blood. If you have an anemic goat that has a negative fecal exam for internal parasites, take a peek under the goat's hair to see

Garden Insecticides

It is not hard to find suggestions on the Internet for using common garden insecticides, such as Sevin, on goats that have external parasites. Sevin was even recommended by vets for use on animals before the United States had legislation regulating off-label use of products. These products have not been tested for safety for use on mammals. There is no research that provides milk or meat withholding times, and the body will absorb chemicals through the skin, which is why some drugs, such as hormone therapy and pain medications, are now delivered through a skin patch. The manufacturers of garden insecticide products specifically state that the products are not to be used on humans or animals. In fact, many of them include warnings to avoid skin contact and to rinse skin for fifteen to twenty minutes if exposure occurs. Many of these insecticides are also known carcinogens.

if you can find any "crawling dandruff." A goat that has a very rough looking coat should also be checked for lice, as should any goat that is rubbing its body against fences and trees excessively. A kid scratching like a dog usually has lice. Lice on goats are host specific and so will not infect humans.

Mites

Several types of mites can infect goats and can cause crusty lesions, hair loss, or mange, as well as itching and scratching. A vet needs to do a skin scraping to get a positive diagnosis, although this is not always accurate. Goats can also get ear mites, which cause discharge and crusty lesions in the ear.

Ticks

Although ticks can infest goats, it is not common in all parts of the country. For example, I have found three or four ticks on my goats in eleven years, even though we have found hundreds of ticks on the humans on our Illinois farm during that same time frame. In contrast, in wooded areas on the Gulf Coast of the United States dozens of ticks on goats have been reported, so this can vary tremendously depending on where you live. If only one goat has with a problem with ticks, there may be a problem with its immune system.

Fleas

According to the textbooks, the species of flea that attaches itself to cats can also feed on goats, although this seems to be very uncommon. I have never found a flea on any of my goats.

Controlling External Parasites

In addition to being used for internal parasites, ivermectin and moxidectin are sometimes used as a pour-on for external parasites, such as lice and mites. Oral and injectable dewormers do not kill biting lice. An all-natural way to eliminate lice during warm weather is to clip all the goat's hair as you would for a show. The size 10 blade on dog clippers works well. The lice will fall off with the hair as you are clipping the goat. Because they are host specific, the lice will die if they are not on a goat.

CHAPTER 7

Injury, Illnesses and Diseases

HAVING GOATS is a lot like having children. Illness and injury are inevitable, and it is easy to get paranoid about every little cough. In reality, goats seldom get sick when they are living in clean conditions and are well nourished. And buying healthy animals that come from herds free of contagious diseases is also key to avoiding disease. However, when a goat does seem to be a little "off," you want to know what's wrong!

Although this section covers the most common illnesses, there are also many other reasons that goats can have problems. For example, an adult goat that is suddenly losing weight may have a dental problem that keeps it from eating enough or chewing its food properly. A kid that doesn't gain weight may have been born with a cleft palate. A goat that suddenly stops eating may have consumed a plastic bag that is causing a blockage. A doe that doesn't get pregnant may actually be a hermaphrodite. Luckily, these problems are rare, and breeders who have had hundreds of kids may have never had any of them.

The purpose of this section is not to replace competent veterinary care but to educate you about when you might need to see a vet. Just as you don't need to run to a doctor every time you sneeze, there are times when you can take care of a sick goat at home. However, if you ever have any doubts about what is wrong with a goat, especially if it is off feed or unable to stand, you should consult a vet. This section is alphabetized to make it easy for you to find things in a hurry.

Abortion and Stillbirth

A spontaneous abortion is a pregnancy that ends naturally before the babies are old enough to survive outside the mother, which is usually around 132 days gestation. There are many reasons why a doe may abort, and you may never know that it has happened if it occurs early in pregnancy. Furthermore, it may be nothing to worry about if it happens only once. However, if many does in your herd abort or one doe continually aborts, it may be an indication of a nutritional deficiency or a disease.

Deficiencies in copper and selenium are the most common nutritional causes of reproductive problems, including abortions and stillbirths, but deficiencies of iodine, calcium, manganese, and vitamin A can also cause abortions, stillbirths, weak kids, or a kid dying within the first couple days after birth. Excessive selenium in the soil and therefore the forage has also been shown to cause abortions.

To protect yourself, assume infection is the cause of a stillbirth during the last two weeks of pregnancy. Wear gloves when handling the fetus and placenta because many of the causes of goat abortions are zoonotic, meaning that humans can become infected. You may want to send the fetus and placenta to a lab for analysis to determine the cause of the stillbirth. It is advisable to keep the doe separated from other does in case she has an infection that could cause other does to abort, resulting in an "abortion storm."

Toxoplasmosis is the most common infectious cause of abortions and stillbirths in goats. It is transmitted by cats, which may appear completely healthy. Toxoplasmosis is contagious to humans and can cause miscarriage in pregnant women. Just as pregnant women are advised not to clean cat litter boxes for fear of infection from the feces, a pregnant woman should not be exposed to an infected goat or to infected barn cats. There is no treatment for toxoplasmosis, which is why some people with goats do not want barn cats in spite of their mousing abilities. Young cats are the most likely to be infected with toxoplasmosis, however. Once infected, a cat develops lifelong immunity, so older cats are usually not as much of a risk to have in the barn.

Chlamydia may also cause spontaneous abortions in goats. It can be treated with tetracycline and other antibiotics. There is also a vaccine available, but because there are many reasons why a goat might abort, a vaccine should be used only after getting a definitive diagnosis on the cause.

Other infectious causes of abortions in goats include brucellosis (also known as Bang's disease or Malta fever), campylobacter, listeriosis, leptospirosis, Q fever, salmonellosis, and tick-borne fever.

Some drugs and toxic plants can cause abortions or birth defects. The dewormer albendazole should not be used during the first three months of pregnancy, and there is anecdotal evidence that levamisole is not safe to use towards the end of pregnancy.[11] Although some stillbirths have been reported after administering levamisole towards the end of pregnancy, researchers have not found a link in controlled studies. Until there is definitive evidence, however, owners may want to use other dewormers during pregnancy.

The issue of toxic plants is a frustrating one for goat owners as lists of toxic plants abound, but there are many anecdotal reports of goats eating many of the plants listed without ill effects. You can't assume that just because a plant, such as yew or oak, is toxic to another species, it will be toxic to goats. I've seen walnut on a few lists, yet our goats often go through a rotation in our walnut grove. Pine trees are also sometimes mentioned as a plant that can cause abortion in goats, yet I know people whose goat pastures are covered with pine trees. If you have someone in your area who has goats, ask if they're aware of any native plants that cause a problem for their goats. Cases of goat poisoning from plants are actually quite rare.

Stress or trauma can also cause a doe to give birth early. For this reason, it is not a great idea to sell a doe during the last six weeks of pregnancy. In addition to the usual stress of moving, there is also the question of how well a goat will fit into a new herd. Of course, you could also have a goat born on your farm that finds itself at the bottom of the pecking order. A yearling born on our farm was picked on mercilessly by the other goats. At the moment we realized she was in labor—two weeks early—another doe was slamming into her belly, which was, unfortunately, not the first time we'd seen another goat do that to her.

Abscesses

Caseous Lymphadenitis, usually called CL, is the most common cause of skin abscesses. CL is highly contagious because it can infect goats through unbroken skin. CL is unique in that it most commonly affects lymph nodes in the neck. The only way to know if a goat has CL is to have a vet aspirate the contents of the swollen area and culture it to see if it is positive for CL. It is a good idea to isolate a goat with an abscess. If the abscess bursts, the pus that drains from the wound will be highly contagious if it is CL. Once a goat is diagnosed with CL, it is positive forever, and it could have internal abscesses. A blood test for CL is also available.

Although a vaccine is available for CL, it is only used in herds that already have an outbreak of the disease, and it is only given to goats that are not already infected. Once a goat is vaccinated with this vaccine, it will test positive, which means that testing becomes a worthless tool in determining which goats are actually infected.

It is very common for goats to develop an abscess at the site of an injection, whether for medication or vaccination, so it is helpful to make a note of the location of the injection. More than a few goat owners have panicked when finding one of those abscesses, worried that the goat has CL. Injection-site abscesses should not be disturbed, and they will go away within a couple of weeks on their own. They can look really dreadful, however, so some breeders who show their goats will give vaccines and other injections under the armpit of a goat that they know will be shown in the near future so that there won't be a big bump over the goat's ribs where everyone can see it.

A salivary cyst is one of the abscesses most commonly confused with CL because it occurs on the head in the same general area as the lymph nodes. Not every swollen spot on a goat is an abscess. Swelling around the lips and cheeks may be due to the goat eating thorny bushes or other plants that caused a minor injury to the skin. The loss of a tooth may cause swelling around the mouth. Goats can also get goiters on the thyroid just like humans who are deficient in iodine, and this will cause swelling in the neck. Bottle jaw, caused by parasites, is another cause of swelling under the jaw.

Acidosis

The pH in the rumen can become imbalanced when too much grain is fed, especially when a goat is not accustomed to grain. A novice owner might unknowingly overfeed grain, or a goat might break into the chicken house and help itself to the chicken feed. As noted earlier, grain is not the ideal feed for ruminants, which is why many goat owners keep baking soda available free choice to serve as a rumen buffer. Throughout the day goats should have access to roughage, such as hay or pasture, which helps to keep the rumen functioning normally. Rather than feeding the goats' full ration of hay first thing in the morning, the hay should be split into several smaller feedings to be given through the day to keep the rumen working continuously. This also helps to reduce hay waste, as goats are more likely to finish each smaller serving.

Bloat

Goats on pasture rarely get bloat, provided that any dietary changes occur gradually. Improper feeding of legumes or grain can cause frothy bloat, which should be considered an emergency as it can cause death within an hour or two. Suddenly grazing lush fields of legumes, such as alfalfa or clover, and sometimes even wet spring grass can cause frothy bloat, which is when a buildup of gas cannot be released from the goat's rumen, putting pressure on the heart and lungs and ultimately causing death. Goats can also get bloat by gorging themselves on grain. Some experts say that bloat can also be caused when goats eat grain that is finely ground.

The goal of treating frothy bloat is to break down the foam, which is usually done by drenching with 100–200 cc of cooking oil or mineral oil. There are also commercial anti-bloating medications available. Although some people recommend putting a stomach tube into a goat to administer the oil, there is no benefit to doing this, and it is in fact much more dangerous than simply drenching a goat because of the risk that an inexperienced person could put the tube into the goat's lung and kill it. If a goat can swallow, it can be given liquids orally with a drenching syringe. If a goat with bloat cannot swallow, a veterinarian should see it as soon as possible. After drenching, the goat should be encouraged to walk around, or the rumen can be massaged to help the goat release the gas.

Brucellosis

The most common symptom is abortion or stillbirth. Luckily, it is rare in goats in North America, but there is a blood test available if you want to be sure your goats don't have it. Consuming raw milk from a goat with brucellosis can cause Malta fever in humans.

Caprine Arthritis Encephalitis

Usually referred to as CAE, caprine arthritis encephalitis is caused by a retrovirus, and there is no cure. It is transmitted through milk, colostrum, and other bodily fluids. Although it used to be extremely common in goats twenty to thirty years ago, it is less common today as a result of breeders testing their herds annually, culling positive animals, and feeding heat-treated colostrum or colostrum substitutes and pasteurized milk to kids of positive does. CAE is debilitating to goats and may eventually cause arthritis, mastitis, pneumonia, and weight loss. In kids under six months of age, CAE can cause a form of encephalitis, which is essentially the same disease as ovine progressive pneumonia (OPP) in sheep. For this reason, sheep should also be tested, and some would say that they should always be isolated from your goats. CAE is not believed to be contagious to humans.

Goats with CAE do not always show symptoms, which is why testing is important. Some goats can be positive for many years without ever appearing sick. Unfortunately, they are contagious and will give the disease to their kids, as well as other goats in the herd that may come in contact with their milk, colostrum, blood, nasal secretions, or vaginal discharge. For example, if a CAE-positive doe gives birth in the pasture, another doe that cleans off the kid could become infected with the CAE virus. A buck with CAE can also transmit the virus to a doe during breeding.[12]

CAE doesn't kill goats directly. However, they can wind up dying from chronic mastitis or pneumonia, and some owners euthanize CAE-positive goats when arthritis makes it difficult for them to walk as the joints in their legs become swollen badly. Weight loss usually occurs because the goat doesn't want to get up and graze or walk to the hayrack or feed pan, but goats may also lose weight even without the development of arthritis.

Although the AGID (agar gel immunodiffusion) test was used for many years to detect CAE, the ELISA (enzyme-linked immunosorbent assay) test is now the preferred test because it is more sensitive in detecting CAE antibodies. The PCR (polymerase chain reaction) test is another option but is much more expensive than the ELISA or AGID tests. PCR tests for the specific virus rather than antibodies, so it may detect the disease in some goats that would test negative with an antibody test. Because most people use ELISA or AGID tests for antibodies and some goats can have the disease for several months before seroconverting—having enough antibodies in their blood to test positive—many breeders with CAE-positive goats in their herd will bottle-feed all kids with pasteurized milk. CAE-positive adults can infect other adults, so goats that are CAE positive and those that are CAE negative should be kept separated. This also means you must be present for every birth so that you can remove kids before they have the opportunity to nurse from a CAE-positive dam.

Cryptosporidiosis

Cryptosporidium is a protozoan parasite, like coccidia, and diarrhea is the main symptom. It is very hard to treat as there are no effective drugs available, and kids sometimes die as a result. While coccidiosis occurs in kids older than three weeks, crypto can infect kids a few days old. In older kids, you will need a fecal exam to determine if the cause of diarrhea is coccidia, cryptosporidia, or something else. Cryptosporidiosis is most often seen in kids that are kept in cramped, unclean quarters indoors.

To control an outbreak, when kids first get diarrhea, they should be isolated from other kids to decrease the risk of others also getting sick.

In dam-raised kids, this may mean moving the dam and siblings, along with the sick kid, to another stall or pen. It's a good idea to wear latex gloves when handling a kid with diarrhea because cryptosporidiosis is zoonotic. Luckily, this disease is extremely rare in healthy herds that are kept in clean conditions.

Enterotoxemia

Sometimes called overeating disease, it progresses so quickly that the first sign of enterotoxemia may be a dead animal. It's a disease of the digestive tract that occurs when a goat has eaten moldy or otherwise contaminated feed or too much feed. Many people think it is the same thing as bloat, even though the two conditions have very different symptoms. You may have heard or read that the CDT vaccine prevents bloat, but it actually prevents enterotoxemia. This confusion probably arose from the fact that accidental overfeeding of grain can cause bloat or enterotoxemia, so some people advocate a CDT vaccine when you discover a goat has overindulged in grain accidentally. The CDT vaccine does nothing to prevent bloat, and it is not terribly effective in preventing enterotoxemia in goats. This disease is much better understood and more researched in sheep and cattle, and the vaccine works better in those species.

Enterotoxemia causes watery, bloody diarrhea in goats. An affected goat is usually lying down and screaming in pain. When you see a goat with enterotoxemia, you will know that you need to call the vet immediately. Treatment includes the administration of an antitoxin for *C. perfringens D* as well as antibiotics and intravenous or subcutaneous fluid administration. Even with veterinary treatment, many goats with enterotoxemia die.

Foot Rot

The name "foot rot" exactly describes this condition. It usually develops when goats are exposed to wet pastures or wet bedding for a prolonged time. The skin starts to break down and eventually gets infected. This condition is less common in goats than in sheep and can usually be prevented by providing goats with dry bedding and trimming hooves regularly, as well as by keeping goats inside when it rains or when pastures flood. Foot rot is contagious, so animal with this condition should be isolated from other goats. Commercial medications for foot rot are available. Some goat owners recommend vaccinating a goat with foot rot for tetanus because the risk of tetanus is greater in a foot with broken skin.

Hardware Disease

This is not really a disease at all, but means that a ruminant has ingested metal. And, of course, the metal can make the goat very sick. It is more common in cows, although goats are more likely to eat plastic bags or baling twine, which also can cause problems. When a goat consumes something that is not digestible, the only symptoms may be lethargy and a loss of appetite. When a goat is not eating, drinking, walking around, or chewing its cud, it is usually time to call the vet.

Hypocalcemia

The terms "milk fever" and "hypocalcemia" are frequently used interchangeably, although technically hypocalcemia is the cause of milk fever. When a doe becomes deficient in calcium during late pregnancy or early lactation, she will go off feed, show signs of mild bloat, become lethargic, and have difficulty standing or walking. If she has not yet kidded, she may have weak or ineffective uterine contractions. If she has already freshened, her production may suddenly decrease. Her body temperature will also go below normal, which makes the term "milk fever" rather confusing.

This is one of those areas where conflicting information abounds, claims that too much calcium in the diet causes it and claims that too little calcium causes it. Because most of our knowledge about lactating animals in North America is based on cows and too much calcium is the cause of milk fever in cows, it was also believed that too much calcium caused milk fever in goats. Recent research in goats, however, has shown that goats do not require the same low-calcium diet that cows do, and in fact, can develop hypocalcemia if they do not get enough calcium-rich feed such as alfalfa.

If you think a doe might be in the early stages of hypocalcemia, you can give her an oral calcium drench made specifically for goats. It is not a good idea to give human calcium supplements because they are not well absorbed. If she does not show improvement within a couple hours, contact your vet, who will probably treat her with an injectable form of calcium. It is easy to mistake other illnesses for milk fever; however, a doe with milk fever will respond within a couple of hours to calcium injections.

Goats with milk fever are especially susceptible to ketosis.

Infertility in Bucks

Luckily, infertility in bucks is rare. An inability to get does pregnant is usually related to nutritional deficiencies, which is why a good mineral is essential for bucks. As a buck gets older, his sperm count may go down, meaning he can service fewer does in a day, which is something to keep in mind if you have two does go into heat at the same time. Older bucks may also start to have problems with arthritis or other aches and pains that make it difficult for them to mount a doe.

If a young buck has been bred to multiple does and never settled any of them, a visit to a veterinarian is in order to make sure he is genetically a buck. An intersex goat may appear to be a buck on the outside, even though lacking all of the necessary anatomy.

Infertility in Does

Based on emails I receive and posts on my online goat forum, it seems that a lot of people worry about the ability of their does to get pregnant. The reality, however, is that less than one percent of does have a genetic inability to get pregnant, so it isn't something that you are likely to experience unless you have a sizeable herd. Freemartins and hermaphrodites are very uncommon in goats, but a quick physical exam by your vet or anyone trained in artificial insemination will tell you if your doe is physically capable of getting pregnant.

The most common reason goats do not get pregnant is a mineral deficiency, and this is fairly common. Copper and selenium both play an important role in a goat's ability to come into heat, get pregnant, and stay pregnant for five months. If several goats in the herd are having fertility problems, you should look at your feeding and supplement program to ensure that it is providing adequate amounts of copper and selenium.

It is possible for does to have cysts on their ovaries, which could keep them from settling, but it is not common. A cystic doe may appear to come into heat every week or not at all. This can be tricky for an inexperienced goat keeper, who may not notice a doe coming into heat if it isn't one of the more vocal ones. A cystic doe can be treated with hormone injections available from your veterinarian.

In does that have freshened before, it is possible that a subclinical uterine infection from the previous delivery is keeping her from getting pregnant or staying pregnant. This is more likely to occur following an assisted delivery where the attendant had their hand inside the uterus. Some people will automatically administer antibiotics to a doe after an assisted delivery with the assumption that it will take care of the risk of infection. A low-grade infection, however, may not have any other outward symptoms, so you would have no idea whether the antibiotics had worked until you found yourself with a doe that wasn't getting pregnant.

Johnes Disease

The possibility of bringing Johnes onto your farm is one reason you should not buy goats from the sale barn. Although Johnes is rare in goats, it also infects cattle and sheep, which means that a goat that was healthy when it arrived at the sale barn could pick up the disease while there. Although Johnes is just one of many diseases that an animal could pick up at a sale barn, it can be one of the most devastating. It is a disease that comes onto your property through the introduction of a new animal that appears to be perfectly healthy. An animal can be carrying Johnes and shedding the virus in feces, contaminating the pasture, before they appear to be sick. Transmission is fecal-oral, meaning that your entire herd could be infected in short order. Johnes can survive on the pasture for several years.

The only way to know if an animal has Johnes before it shows symptoms is to test. There is no test for Johnes that is extremely accurate in detecting infected animals, which means that it is not terribly informative to have a single negative test result on a single animal. It is more reassuring to have a whole herd test negative, and it is even more reassuring to have annual negative whole herd tests. After several years of negative results in a closed herd, the odds of Johnes in that herd are as close to zero as one can get. Because Johnes is so contagious, more than one animal in a herd will have it, so odds are much better that there will be some animals testing positive in the whole herd test if the disease exists in that herd.

The most common age for infection to occur is in the first month or two of a goat's life, although they won't develop symptoms for a couple of years. Older goats that are exposed to Johnes may not contract it.

Weight loss is usually the only symptom of Johnes in goats, but weight loss is also associated with parasites, dental issues, and other diseases. Even social hierarchy within the herd can mean one goat isn't getting its fair share of hay during winter when pastures are dead. Bucks also tend to lose a lot of weight during breeding season, sometimes as much as 20 to 30 percent of their normal body weight. Because copper deficient goats tend to have poor parasite resistance, their body condition may be poor even when they have a small parasite load, which leads some owners to worry that their goat has Johnes.

There is no vaccine and no cure for Johnes, so if you have an animal with the disease, euthanasia may be the best solution.

Ketosis

Also known as pregnancy toxemia, ketosis usually occurs in goats in late pregnancy or early lactation when they are undernourished, although it can also occur in obese does. It is most common in goats that are carrying three or more fetuses or who are especially heavy milk producers. Other than going off feed, symptoms of ketosis during pregnancy are not obvious, although the doe's breath may smell sweeter than normal. If in doubt, compare her breath to other does in the herd. A doe in milk with ketosis will have a decrease in production.

There are a couple of options available for detecting ketones in a goat's urine, and one that will detect it in milk or blood. Ketone test strips or powder are available from goat supply companies online and from farm supply stores. If a goat tests positive, she can be treated by dosing with propylene glycol or one of the liquid supplements that has propylene glycol as the first ingredient, until testing is negative for ketones in the urine.

Listeriosis

This is an infectious disease caused by the listeria bacteria, which can survive in the environment for years. The disease can lie dormant in an infected animal until the animal becomes stressed. A goat with listeriosis does not necessarily walk around in circles, in spite of its common name, circling disease. An infected goat may seem uncoordinated, not want to stand, hold its head twisted to one side, stop eating, and lose muscle control of its face so that its ears and eyes are droopy and its tongue hangs out. You should seek veterinary attention as soon as possible. Goats with listeriosis can die quickly if not treated.

A goat with listeriosis will shed the organism in its milk, and it can be transmitted to humans if the milk is consumed raw.

Mastitis

Mastitis is an infection of the mammary gland, and it can range from sub-clinical to fatal. Unless you are on milk test or test your goat's milk yourself, you will have no idea that your goat has a subclinical infection. The commercial dairy industry tests cows regularly for mastitis because the only real symptom of sub-clinical mastitis is lowered production. The reduction can be so minimal that you wouldn't notice it in a home milk goat, although when added together, this loss of production in commercial dairy animals adds up to millions of dollars a year in lost profits. Sub-clinical mastitis may or may not develop into full-blown mastitis.

If you are on DHI milk test, one thing tested monthly is the somatic cell count (SCC) in the milk. An infected doe will shed white cells in her milk. Deciphering SCCs in goats is a little more complicated than in cows because goats may also shed epithelial cells in their milk, so they are notorious for having a higher average SCC than cows. Most of the DHI testing labs in the United States do not use a test that distinguishes the two types of cells, epithelial and somatic, which may result in SCC readings that are incorrectly high. If you are on milk test, ask the lab if they distinguish between the two. If not, you will usually need to ignore their definition of a mastitic goat, which is based on normal cow scores, and figure out your own definition of normal for your individual does. Some does always have higher counts than others, but a sudden increase in a doe's score from one month to the next could be a sign of mastitis. A gradual increase throughout lactation is usually nothing to be concerned about.

You can easily test a goat's milk by using the California Mastitis Test (CMT) every month. If you are on milk test and your goat has a SCC result of 500,000 or more, it is a good idea to test each half of her udder using the CMT. As you are starting to milk each goat, put a few squirts into your strip cup, then squirt milk directly into a section of the CMT paddle, add the diluted reagent, and swirl. If the mixture remains completely liquid, it means the SCC is less than 200,000, and the goat does not have mastitis.

When I have a goat that has a high SCC or a CMT that shows the possibility of mastitis but she is not otherwise showing symptoms of

---

❖ **GIANACLIS CALDWELL, Pholia Farms, Rogue River, Oregon**

So what do we do when one side of the udder has an obvious problem, confirmed by CMT (California Mastitis Test)? First you must rule out problems with milking equipment and general health of the animal. Of course, when it is just on one side, you have to assume an udder infection of some sort. Before you resort to antibiotic usage, you can try some organic and old-fashioned remedies. I used to do peppermint oil rubs to the udder and give the doe an oral dose (about 60 ml) of her own milk to hopefully stimulate an antibody response. I have recently added a certified organic producer's common technique of orally dosing the animal with garlic "tea." What a miracle it has been! We soaked peeled garlic cloves in water (be sure to keep it refrigerated as botulism is a risk if not). Then we dosed the doe with 40–60 ml three times a day, and her SCC went from scores of 722,000 and 652,000 to—are you ready?—1,000. Yup. Garlic. Keep in mind that many high somatic cell counts and cases of subclinical mastitis will spontaneously disappear without garlic, but it's just another tool that may be helpful.

mastitis, I will try some alternative therapies such as peppermint and oregano essential oil diluted in a carrier oil massaged on the udder. Most organic dairies use peppermint oil as their first line of defense against mastitis. I added the oregano after hearing a botanist talk about it being used for mastitis in women. Some people also use tea tree oil. If you are not experienced in using essential oils therapeutically, there are commercially available udder balms that include these ingredients.

If a goat has a full-blown case of mastitis, she will have a hard udder that is very warm to touch. The milk will often taste salty. If nothing comes out when you start to milk her, examine the teat. There might be a plug of milk in the tip of the teat, and a long string of milk may come out when you pull on the plug. Although it may be difficult to get milk out of an inflamed udder, it is important to do it. All instructions for treating mastitis start with milking. Holding hot compresses on the udder while milking is helpful, but of course, you don't have three hands, so it's easier to do when another person can hold the compresses while you milk. Milking the doe four times a day will help. Although there are over-the-counter antibiotics that are put directly into the teat of the affected half, it is a good idea to call your vet and have the milk cultured so that you use the correct antibiotic. You should see improvement in two to three days if you're using an antibiotic that is appropriate. Unfortunately, antibiotic resistant mastitis is becoming more common, which is why it is important to avoid using antibiotics as much as possible. Unresolved mastitis can result in a ruptured udder or in gangrene in the affected half of the udder. In severe cases, the doe can die.

Mineral Deficiencies

If goats don't get enough of any vitamin or mineral, deficiencies will obviously occur. However, some deficiencies are more common than others and will cause more severe problems.

Copper

Copper deficiency is far more common than most people realize. It is more common today than in previous years because much of the soil in the United States is deficient in copper, which means plants grown in it are deficient. A prevalent symptom of copper deficiency in my herd is a faded coat because copper is responsible for pigmentation of hair. My gold goats turn cream, cream goats turn white, and black goats turn rusty red. They also lose hair on their faces. A fishtail or bald tail tip is another common symptom. Severe orthopedic problems, such as bent legs, swaybacks, or spinal injuries, can be caused by copper deficiency. I have not seen deformed legs or swaybacks in my herd, even though I've had several goats die from confirmed copper deficiency, so don't assume that your goats have enough copper simply because they all have straight legs and spines.

What I learned from Muse

We'd had Nigerians for three years when I purchased Muse, my first LaMancha, a yearling in milk. At six months fresh, she dried up, which was disappointing, but I was eagerly looking forward to her kidding the next spring. We never saw her come into heat until December, even though we had two wethers with our does to help detect heat cycles. Because I didn't own a LaMancha buck at the time, I drove her to another farm seventy-five minutes away for a driveway breeding. In spite of the buck successfully covering her twice, she came back into heat three weeks later. Not being in a position to drive to the other farm again, I decided to breed her to my Nigerian buck. She never got pregnant that year, but she wasn't the only goat having fertility issues. We had about a dozen Nigerian does at that time, and several were not coming into heat or getting pregnant. My daughter Margaret did some reading and said she thought our goats were copper deficient, but over the months, four different vets said that was impossible.

A year after we bought Muse, I bought a LaMancha buck, and the following spring Muse kidded with twin does. By June, however, she still had not shed her winter coat when we clipped her for a show. A couple of weeks later, she died unexpectedly. Having no idea what was wrong with her, I took her body to a vet for a necropsy, and I said I wanted to have her liver checked for copper. The necropsy came back listing Tyzzer's disease as the cause of death, and her copper level was 4.8 on a scale where normal is 25–150 ppm.

Copper is also very important in reproduction, so deficient goats may not come into heat or may have difficulty staying pregnant. In my experience, bucks seem to have more problems with copper deficiency than does, probably because the does are getting a commercial goat ration that is fortified with 40 ppm copper. My bucks are consuming only pasture and hay, which have much lower amounts of copper, so they need to consume enough of the goat minerals as a supplement to get an adequate amount of copper.

Copper deficiency can be primary or secondary. Primary copper deficiency means that the goats are not consuming enough copper. In

Tyzzer's is a rodent disease, and in searching scientific journals, I was not able to find a single case in a goat recorded, although there have been a few cases in horses. Like any barn, ours has mice, and Muse probably ingested mouse poop at some point. But why would she get a disease that is unheard of in goats? My assumption is simply that her immune system was not functioning at an optimum level, leaving her vulnerable to a disease that a healthy goat would never have contracted. The vet insisted the whole thing was just a fluke and said that there was nothing I could have done to prevent her death, and he refused to give me prescription copper for my other goats.

I started doing a lot of reading, however, and realized we had the worst possible scenario for creating copper deficiency. Our well water had sulfur in it, which is a copper antagonist and reduces absorption of available copper in the diet. On top of that, we had been feeding a commercial goat ration with only 10–15 ppm copper. I contacted an animal scientist who had authored published articles on goat nutrition, and he advised finding a feed with 35–40 ppm copper. We also began giving the goats COWP.

After changing our goat feed and giving COWP to the goats, we immediately saw faded, wiry-haired goats shedding their coats and replacing them with much softer and darker hair. That fall, all of the does came into heat, were bred, and stayed pregnant until term.

secondary copper deficiency, the goats are consuming plenty of copper, but antagonists in the diet, such as high molybdenum, sulfur, or iron, interfere with the goat's ability to absorb and use the copper. For example, alfalfa can have levels of molybdenum that are so high as to induce copper deficiency in goats. Unfortunately, the level of molybdenum varies tremendously from one location to another, making it impossible to say whether or not it is causing a problem in a particular herd unless you have your hay tested. Well water that is high in sulfur or iron can also cause copper deficiency. Excessive iron in the water generally turns white sinks orange, and excessive sulfur makes the water stink like a dirty dishrag or rotten eggs.

Copper deficiency is a challenge to diagnose because hair and blood tests are not very accurate. The most accurate test for copper is a liver biopsy, which is not practical in a live goat. I have learned always to ask for a copper level on a goat's liver when a necropsy is done. Mineral levels on livers are not performed routinely when doing a necropsy unless something jumps out at as a potential problem when the vet is discussing your feeding practices. I have removed a liver myself and sent it to a lab for copper testing when I suspected the dead goat had been copper deficient based on a faded coat and other symptoms.

Selenium

Like copper, selenium is an important mineral for the health of the reproductive system of does, and an inability to get pregnant and stay pregnant is sometimes caused by a deficiency of this mineral. A doe with selenium deficiency may have trouble giving birth, is more likely to have a retained placenta, and will have lower milk production than a doe with adequate selenium in her diet. A buck may be less likely to get does pregnant. Kids born to selenium-deficient does are more likely to be stillborn or weak and may suffer from white muscle disease in the first few weeks of life, which can result in death. It is important to understand that the occasional weak kid is probably not selenium deficient. A herd with a selenium deficiency problem will likely see many of these problems in a number of goats.

Soils in North America vary from deficient to toxic in their levels of selenium. In many parts of the United States, soils are deficient in selenium, making it necessary to use fortified feeds as well as free choice minerals that include enough selenium. Although injectable selenium is available by prescription from a vet, some goats have died from toxicity as a result of the injection. It is important that you have evidence of selenium deficiency before using injectable minerals.

When goats die on my farm, I have their livers tested for selenium (as well as other minerals). Years ago they tested at the very low end of the normal range for selenium. Having this knowledge, I gave injectable selenium to my goats prior to breeding season for several years. However, I have switched to a free-choice selenium supplement for a couple of reasons. First, many grass-based cattle producers insist that "cafeteria-style minerals" are the best way to provide minerals for animals. Each individual mineral is available to the animals, and they can take as little or as much as they want. The body absorbs minerals consumed orally better than those injected. When supplements are injected, the majority of the supplement leaves the body in the urine over the subsequent twenty-four hours, suggesting that injection of a routine supplement is not useful in a situation where chronic deficiency is a real problem.

Zinc

The most obvious symptoms of zinc deficiency are in the hair and skin. A goat with zinc deficiency looks scruffy with flaky dandruff and odd shedding at unusual times of the year. I had a buck that developed bald patches of dry skin in the midst of a winter when temperatures were falling below zero Fahrenheit. Another symptom is excessive salivation, which generally makes a buck look like he is foaming at the mouth.

Zinc deficiency is often caused by excessive calcium in the diet, either from mineral supplements or calcium-rich foods, such as alfalfa. Injectable zinc supplements are available by prescription, but they are formulated with other minerals. The combination of minerals might have the unintended consequence of delivering a toxic dose of a mineral the goat does not need. Zinc is also available as a single mineral supplement, but reducing the amount of calcium in the diet may also correct the problem. Some goats seem to be more prone to zinc deficiency than others.

Neonatal Mortality

There are dozens of possible causes of death in newborns. If it happens rarely, it is probably not something that you can remedy. A very small percentage of kids are put together without all their pieces in the right places, and there isn't always anything that can be done about it. However, if several kids in your herd die around the same time, an infection could be causing the problem. Several kids dying at birth throughout the kidding season could indicate a mineral deficiency.

Hypothermia is probably the most common cause of newborn deaths in northern climates during the winter. When the air temperature is below freezing, kids can get hypothermia within fifteen minutes of birth if they are not dried off quickly enough. I've even seen kids get hypothermia when born outside in 45°F weather when it was windy, lowering the temperature with the wind chill.

If several kids die at birth within a few days, having a necropsy performed will determine if the cause of death was an infection or something else that can be remedied. The placenta should be submitted along with the fetus because sometimes the answer is in the placenta rather than in the kid.

Pinkeye

Conjunctivitis in goats does not actually turn the eye red as it does in humans. Instead, the eyeball usually turns white or cloudy blue. Flies that feast on manure and then get into the goat's eyes spread the infection, so fly problems in the barn area should not be ignored. Dust in the eye from eating dusty hay or a poke in the eye by a blade of grass can also cause an eye infection. In severe cases goats can be blind from the infection, although this is usually temporary. It is common practice to isolate goats with conjunctivitis so they don't infect other goats. Because the infection can be spread by flies, however, the effectiveness of separating infected goats will be determined by how far away you are able to move them. A separate building or a distant pasture with a small fly population would provide better protection for uninfected goats than simply putting an infected goat in a separate stall in the same barn as the others.

A variety of over-the-counter medications are available for treating conjunctivitis in goats. However, if pinkeye is just one of many symptoms in your goats, you should see a veterinarian because mycoplasma, chlamydia, listeria, and organisms that cause other systemic diseases can also cause conjunctivitis.

Polio (Polioencephalomalacia)

Goat breeders usually use "goat polio" and "thiamine deficiency" interchangeably, but they are not the same thing. For years, the two terms were used interchangeably because goats with symptoms of polio responded to treatment with thiamine, leading people to believe thiamine deficiency was the cause of the symptoms. There are a number causes for goat polio, however, such as lead poisoning, sulfur poisoning, salt poisoning, moldy hay, too much grain, and not enough water.

Thiamine (vitamin B1) is produced in a healthy rumen, so it is not a vitamin that goats need to consume. Thiamine deficiency can happen whenever the rumen is upset by any number of things, including ingesting excessive grain, which is why it is most often seen in feedlot cattle and sheep. It can happen on the homestead, however, when a goat gets into the chicken grain one time too many. Administration of white dewormers, levamisole, or amprolium can also upset the rumen and cause a thiamine deficiency, especially when used long term.

Stargazing is the most often cited symptom of thiamine deficiency, and I've seen more than a few new goat breeders start giving their goat vitamin B injections for days or even weeks simply because a goat was tipping its head back and moving it from side to side as if it was looking at the sky.

Polio is a disease of the brain, however, so there will be multiple symptoms, and the goat will be very sick. It will be depressed, off feed, and often have diarrhea. Unfortunately, these symptoms are very similar to enterotoxemia and listeriosis. And as the list of possible symptoms gets longer, it just gets more confusing. A goat with polio may also be blind or start circling, which are also symptoms of other diseases. As you have probably realized, a quick trip to the vet is your best bet.

A goat with polio can die within a day or two if left untreated. Because diagnosis is so challenging and treatment for polio is most likely to be effective if started early, the vet usually gives an injection of thiamine, which is by prescription, if polio is suspected. If treatment is going to be effective, it will work within a day or two. In the worst cases of polio, treatment may save the goat's life, but it will never completely recover and will be partially blind or mentally impaired forever.

So, why is a goat with polio treated with thiamine if it isn't the same thing as thiamine deficiency? It often works to reverse symptoms, although no one knows exactly why. Studies have shown that a goat suffering from lead poisoning or sulfur toxicity will respond positively to treatment with thiamine, even though it is not thiamine deficient. This is important to know because the treatment of a goat with thiamine injections is not supposed to be long term. If the goat continues to relapse or if other goats are having the same symptoms, there is something in the diet or the environment that is causing the problem, such as a diet with too much grain, which upsets the rumen balance. You may also want to test your water for the presence of lead or sulfur.

Respiratory Conditions

When a goat has a runny nose or cough, people often assume it has a respiratory infection or lungworms. But those symptoms could be caused by something as simple as dust from hay or living on a gravel road. Ammonia buildup in a barn, as well as smoke or exhaust from machinery, can also cause coughing and runny noses. Treating respiratory issues caused by environmental issues will be a waste of time, as they will not go away until the environmental problem is corrected.

A goat can develop a runny nose following any type of injury to the bones in the head, such as disbudding, a damaged horn, an infected tooth, or a cracked bone in the face. Something as simple as a tight collar can cause coughing. If a goat starts to cough only when being led by the collar, the problem is not illness in the goat. Holding a collar too tightly can restrict a goat's airflow resulting in the goat falling to its knees. This is not an uncommon sight at goat shows when someone is showing a goat that hasn't been trained to lead or simply does not like to lead.

There are many types of pneumonia in goats, and in spite of how common it is, it can be a challenge to diagnose and treat. It can be caused by parasites, fungus, and a long list of viruses and bacteria. Treatment with antibiotics may or may not be effective, depending on what is causing the pneumonia. Sometimes pneumonia can be a symptom of a much larger problem, especially when occurring with stillbirths and abortions.

A kid with pneumonia may cough or have a runny nose, but the only symptoms in adults may be lethargy and going off feed. You may hear rattling in the lungs or base of the throat by using a stethoscope. A

What I learned from Charlotte

During kidding season one year, I walked into the barn and saw Charlotte lying in the straw with her head up. She was alert and chewing her cud, looking perfectly normal for a goat that should be giving birth within a few days. But even though I was several feet away from her, I could hear her labored breathing, which reminded me of Darth Vader. I panicked, immediately thinking she had pneumonia, and I rushed back into the house to pull out all my books and search online for information on pneumonia in pregnant goats. Finding nothing that specifically addressed that topic, I decided to go back to the barn and watch her for a while. A little later when she decided to stand up, the labored breathing disappeared. Charlotte's belly was extremely wide, and the obvious finally occurred to me. She was having difficulty breathing when she lay down because the kids were compromising her lung capacity. A few days later, she gave birth to quadruplets, and my suspicions were confirmed when her breathing was again quiet.

stethoscope is inexpensive to purchase and can be used to familiarize yourself with the sound of healthy goat lungs for comparison. Although a goat with pneumonia will usually have an elevated fever, a temperature below normal is even more of a concern because that means its body is starting to shut down and it is near death.

Inhalation pneumonia will appear in a goat after being drenched with medication that accidentally goes into the windpipe instead of down the throat. It may also happen when a goat throws up, which they do so rarely that some sources say goats don't vomit. However, when a goat does vomit, it is usually because it has consumed something poisonous. A kid born with a cleft palate can develop pneumonia from aspirating milk.

Determining the cause of a respiratory ailment can be tricky, and if the goat has a fever of 104°F or more, is off feed, or lethargic, call the vet. If antibiotics were recommended and you have been treating a goat for forty-eight hours and are not seeing improvement, the antibiotic is not working. You might need a different antibiotic, or the cause of the pneumonia is a virus, which won't respond to antibiotics. If you have seen an improvement, it is recommended that you continue treatment for at least forty-eight hours after symptoms have disappeared. Stopping antibiotic therapy too early contributes to the development of antibiotic-resistant bugs.

There are a number of different organisms that can cause pneumonia, and there are only vaccines available for a couple of them. Even if you vaccinate for pneumonia, you still need to make sure your goats get plenty of fresh air and do as much as possible to create an environment that is not conducive to respiratory infections.

Ringworm

A goat that is losing hair in patches and exposing scaly, crusty skin, especially on the face, neck, and legs, might have ringworm, which is contagious to humans, and should be diagnosed by a vet because it looks similar to mange to the untrained eye. Ringworm is a fungus infection and usually infects goats that are already debilitated by another illness or have poor nutrition.

Scours

Scours is the word used for diarrhea in livestock, although some people treat it as a disease itself. Diarrhea is a symptom. In *Diseases of the Goat*, John Matthews presents thirty-one pages on the causes of diarrhea, which include bacteria, viruses, protozoa, and parasites, as well as nutrition, stress, and toxic agents, such as drugs, minerals, and plants.[13] Treating diarrhea without knowing what is causing it can do more harm than good. Giving a dose of medicine to stop diarrhea in a goat with coccidiosis or an infection will do nothing to cure the actual illness. The danger of giving antidiarrheal drugs is that you may think the problem is gone when in reality you have only masked the symptom, and you could wind up with a dead goat.

The most common cause of diarrhea on our farm for the first several years was the goats breaking into the chicken house and gorging on grain. Luckily the diarrhea was always short-lived and cleared up on its own within twelve to twenty-four hours. It could have ended badly, however, if a goat's rumen had been so upset as to lead to bloat, enterotoxemia, or goat polio.

The most common cause of diarrhea in kids three weeks of age or older is coccidia. Diarrhea in kids younger than three weeks of age is likely due to nutritional problems, such as too much milk or switching from goat milk to milk replacer. An infectious agent, such as *E. coli*, cryptosporidium, or salmonella, may also be the cause of diarrhea in kids under a month of age. These infections need to be diagnosed by a vet.

If one of my goats has diarrhea but no other symptoms and otherwise appears healthy, I will wait a few hours before doing anything because many times the diarrhea goes away on its own, which generally means that the goat ate something that upset its digestive system. Once a goat has diarrhea, its back end is messy, so it is not immediately clear when the diarrhea has ended. I press a paper towel against the messy area to see if it is still wet. If it is dry, it means it has been at least a few hours since the goat had diarrhea, and I'll continue to check to make sure it stays dry. If it is wet, I'll continue checking every few hours, and if it hasn't dried up in about twelve hours, I'll usually treat for coccidia if it is a kid over a month old. Because adults rarely get diarrhea and the cause tends to be something more complicated, I usually check the goat's temperature, listen for rumen sounds and then call the vet.

Scrapie

In the same family as mad cow disease, scrapie is a transmissible spongiform encephalopathy (TSE). At this time, there is no evidence that humans can contract scrapie from infected goats. It is caused by a prion, rather than by a bacteria or virus. There is no treatment, and it is highly infectious. Scrapie is practically non-existent in goats in North America, but because the disease is closely related to bovine spongiform encephalopathy (mad cow disease), complete eradication is the goal of most governments. There is a scrapie eradication program in the United States, and the USDA requires goats being sold across state lines or being exhibited in fairs and sales to come from a herd with a scrapie identification number. Goats must have a permanent form of identification, which is usually a tattoo for dairy goats or an ear tag for meat goats, so that they can be traced back to their herd of origin.

Goats can have TSE for a couple of years before developing symptoms, which include scratching and rubbing on fencing and housing, as well as biting at its legs and sides. A goat may lose weight and have neurological symptoms, such as tremors, incoordination, excessive salivation, and blindness.

Tests are available for the presence of TSE.

Sore Mouth

Contagious ecthyma, also known as orf or sore mouth, infects sheep and camelids as well as goats. The main symptom is a mouth covered in crusty sores, which are highly contagious to other small ruminants and humans. The sores are painful, so an infected goat might not eat much, which will lead to weight loss, and kids may even die when infected. Goats may also get sores on their nose, eyes, ears, anus, vulva, scrotum, and teats. Pus and scabs are highly contagious and can contaminate a pasture for years. Once an outbreak starts in a herd, it usually winds up infecting all goats, and if an individual goat does not get open sores on the mouth, it is possible that it is sub-clinically infected. A sub-clinically infected animal is infectious and can infect other animals if sold. People who handle infected goats should always wear rubber gloves because of the risk of infection to humans.

No treatment is available, and it usually takes a month or longer for the sores to heal. There is a vaccine, but it is a live virus, which will cause a crusty sore at the site of the vaccine, and goats should be considered carriers for several weeks after vaccination. For this reason, the vaccine is seldom used in herds that do not have a history of sore mouth. Unlike the CL vaccine, it can be given to goats that have the disease.

Tetanus

Goats are infected by tetanus through an open wound, and preventing wounds is not always easy, especially when they can occur through regular goat maintenance, such as hoof trimming.

The main symptom of tetanus in goats is muscle rigidity. If the animal is able to stand, it will have all four legs spread far apart like a rocking horse. Once it goes down, it will have its legs and neck extended rigidly. A vet should be called immediately because the disease can progress very rapidly.

Treatment for tetanus is often unsuccessful, so prevention is key. For example, avoid castration methods that break the skin, including the use of elastrator bands, which create an anaerobic environment under the band while the scrotum is dying and preparing to fall off. A vaccine is available, which must be given annually for continued protection, and tetanus antitoxin is available to treat animals that have recently been injured, regardless of whether they have been previously vaccinated. Most veterinary texts recommend cleaning all wounds and flushing with hydrogen peroxide, as well as leaving the wound open to air, rather than bandaging it. Some also recommend applying an antibiotic locally.

Tuberculosis

Tuberculosis is extremely rare in goats, and most states are certified TB-free. Humans can contract TB through drinking unpasteurized milk, so it is a good idea to have your goats tested if you have any doubts about their TB status. This is a chronic wasting disease in goats and will cause poor milk production, diarrhea, and weight loss along with a chronic cough. There is no treatment for livestock with TB.

Urinary Stones (Urinary Calculi)

The male goat has a very small urethra, which is the tube through which urine leaves the body. Research in cattle shows that early castration causes the urethra to be even smaller. Because a goat's penis is so small, a very tiny urinary stone can cause a blockage. A buck standing in the usual position to pee and unable to produce urine may have a stone blocking the urethra. Without treatment the bladder or urethra can rupture within a day or two and cause death. Contact the vet as soon as you realize the buck is not able to pass urine normally. Treatment with ammonium chloride may work if it is started soon enough. Otherwise, a variety of surgical options are available.

Minerals or feed with added ammonium chloride can be used for prevention, but diet is an important contributor to this problem. Urinary calculi are most often seen in bucks or wethers that are on diets with a large amount of grain. Bucks do not normally need grain, and because wethers have an even smaller urethra, they should definitely not receive grain.

White Muscle Disease

The name white muscle disease is often used synonymously with selenium deficiency in young kids. Selenium deficient kids are weak and often die. See the section on selenium under the discussion of mineral deficiencies for more information.

Vaccines

The decision about whether or not to vaccinate your goats is a very personal one. I firmly believe that you should make the decision that will help you sleep at night. Neither choice is risk free. Some people would feel terrible if they had a goat die of a disease for which a vaccine exists, while others would feel worse if a healthy goat died from a reaction to a vaccine.

The CDT vaccine is for *Clostridium perfringens* types C and D, which causes the disease enterotoxemia, and tetanus. It is the only vaccine routinely recommended for goats and is available without a prescription in most farm supply stores and online. Some goat keepers continue vaccinating for as long as they have goats; others decide to stop for a variety of reasons specific to their farm or their philosophy.

Anaphylactic shock is a possibility following any vaccine, so it is a good idea to have injectable adrenaline available, which you can purchase from a vet.

It is not unheard of for a vaccinated goat to get enterotoxemia, which has led some breeders to vaccinate every goat two or even three times per year, although the manufacturer does not recommend this. Many people also will vaccinate a few goats today, a few more next week, and a few more next month using the same bottle, even though the manufacturer instructions say that the entire contents should be used when the bottle is first opened. It seems prudent that if you are going to use a vaccine, you should follow the manufacturer's recommendation for best results.

For those who choose to vaccinate with CDT, kids usually receive their first injection at five to six weeks of age and then get a second shot three to four weeks later. Annual booster shots are recommended. Does are usually vaccinated about a month before kidding so that the antibodies are passed on to the kids before birth. You should know a goat's vaccination status when you purchase it if you plan to continue vaccinating. Annual injections will not be very effective if the goat did not receive the first two shots three to four weeks apart.

Vaccines are available for abortions, CL, pneumonia, rabies, and sore mouth. It is best to consult with a vet to determine if there is a need to use any of these vaccines in your herd. There isn't a rabies vaccine approved for use in goats, which means your vet will have to use a vaccine created for a different species.

CHAPTER 8

Breeding

✦ ✦ ✦

IF YOU ARE RAISING your goats to produce your own milk and meat, they need to be bred regularly. If you decide to breed goats of different breeds, ideally their mature size should be similar. When breeding two goats from breeds that are different sizes, the buck should always be from the smaller breed so that you don't encounter difficulties during kidding. The average Nigerian Dwarf kid is about 3 pounds at birth, while the average standard-sized kid can easily be 7 or 8 pounds, so it is easy to see that you could wind up with a caesarean section on a doe that was bred by a buck from a larger breed.

Breeding Season

It is common practice to breed goats to kid annually. For example, a doe kids in January and starts producing milk. Seven months later, in August, she is bred so that she will freshen again in January. Milking is stopped when she is three months pregnant so that her body can concentrate on putting all of its energy into the growing kids during the last two months of pregnancy. You can see that if you stagger breeding by three or four months, you will never be without fresh milk.

Unlike most livestock, which can get pregnant year-round, goats are seasonal breeders, meaning they get pregnant only in the late summer or fall for late winter or spring kidding. Although some sources say that Nigerian Dwarf goats can be bred year-round, not every Nigerian will come into heat in the spring. If you have only a couple of does and want milk year-round, you might plan to have one kid in January and one in May.

A doe must be in standing heat to get pregnant, which means she will stand and let a buck breed her. If she is not in heat, she'll run away when he tries to mount her. Does come into heat about every twenty-one days, plus or minus two. If you see a doe in heat but you are not ready to breed her, mark your calendar so that you'll know to pay close attention to her behavior in three weeks so that you don't miss her next heat cycle.

Think about where you will be in five months when your doe comes into heat. If you are planning a vacation for that time, it's probably a good idea to postpone breeding until the doe's next heat cycle. A breeder who lives in a cold climate and has a full-time job off the farm might choose to breed for a summer due date so that they don't have to worry about hypothermia if they are not there when kids are born.

Breeding Age

A lot of people wonder how old a doe should be when she is first bred. However, size is more important than age. Although some people think a doe under a year of age may not be a good mother, this varies from goat to goat. A goat that is too small, though, will not be able to give birth. It is generally recommended that does be at 60 to 70 percent of their adult weight before being bred. Some does reach that size by the time they're seven months old, although it can take up to eighteen months for others. On the rare occasion that a doe doesn't reach that size by age two, she should not be bred. Breeding one doe that was too small was more than enough to make me adopt a lifelong better-safe-than-sorry policy when it comes to breeding does.

Sexual maturity with bucks is individualized. There are anecdotal reports of two-month-old bucks breeding does, but there are six-month old-bucks who don't know which end of the doe to mount. Just because a buckling is mounting doelings does not mean that he is actually

What I learned from Giselle

When I started raising goats, everyone I knew talked about breeding does to kid as yearlings. It never crossed my mind when I bred Giselle at thirteen months that she might not be big enough to give birth safely in five months. All of my other young does had been bred to kid by the time they were her age, and none had had any problems.

Five months later, it was Christmas Eve and Giselle was in labor. I could see two hooves, but regardless of how many times she pushed, the kid wasn't emerging. I pulled gently on the two legs and made no progress. I pulled harder and still made no progress. One of my daughters tried, and we agreed the kid was stuck. We'd heard about people pulling out a goat's uterus, so we didn't want to pull too hard. I called the vet, and he said he'd meet us at his office. We put Giselle on the front seat of the pickup truck between us. It took almost an hour to drive to the vet, but Giselle was quiet because she was already exhausted.

I assumed Giselle would be having a caesarean, but when we got to the vet's office in the early morning hours of December 25th, he said he could probably get the kid out vaginally. He attached a leg snare to the kid's front hooves and pulled so hard that he would have pulled Giselle off the table if my daughter and I had not been holding her. After a great deal of pulling and even discussing the possibility of damaging the kid's head, he finally pulled the single kid out. Giselle was so stressed that she completely ignored the kid and just stared into space. On the drive home, she continued to ignore him, and we realized we would need to bottle-feed the buckling.

A few days later I called one of my mentors and told her about the experience. When I told her that Giselle, a Nigerian Dwarf, weighed 35 pounds at the vet's office, she said that she didn't normally breed goats that small. For months I said I'd never breed Giselle again, assuming that she just had an unusually small pelvis, but I finally did breed her more than a year later. Since then she has given birth without difficulty to kids that were more than 4 pounds, which is above average for the breed. And since then, I don't breed Nigerian does until they weigh at least 40 pounds.

producing sperm, and everyone I know who has tried to use a two-month-old buck for breeding has failed. If it is important that your does get pregnant on a certain schedule, it's a good idea to use bucks that are at least six to eight months old.

Signs of Estrus

- **Flagging:** The doe will be wagging her tail a lot.
- **Mounting:** A doe in heat may mount other does or let them mount her.
- **Vocalizing:** Some does won't vocalize at all, but others will be screaming so much that anyone within a quarter of a mile will know she's in heat.
- **Flirting:** A doe will hang out next to a fence where she sees a buck. Even if there are two fences between them, she will get as close as possible.
- **Discharge:** A clear or white discharge might be visible on the tip of the vulva. The color of the discharge is a clue about what stage of heat the doe is in. If it is clear, she is usually in the early stages, but if it is white, she is near the end.
- **Drop in milk production:** This is temporary and will last one or two milkings. It may be a minor drop or substantial.

Breeding Methods

There are benefits and drawbacks to both pen breeding and hand breeding, and the decision to do one or the other is usually based on what is most convenient for the owner. While pen breeding is more convenient at breeding time, hand breeding usually provides a smaller window for the due date.

Those who breed goats primarily for meat usually pen breed, which means they will put a buck with a group of does and let them run around together for three or four weeks, during which time the does should all come into heat and get bred. If you are milking the does, be aware that the milk might be tainted with a goaty taste that results from bucks, which are stinky, rubbing on the does. However, unless a doe is in heat, most won't let bucks do this, so you might have only one day of milk that you don't want to use for human consumption. You can feed it to the chickens, pigs, or barn cats, and they won't mind the taste.

Although I have done pen breedings with goats that are not in milk, such as first fresheners, I am not a fan of this practice because in most cases the due date is a mystery. When goats are together twenty-four hours a day, the odds are pretty slim that a breeding will occur when you happen to be observing. Goats usually give birth about 147 to 152 days after being bred, but it can be a little earlier or later. If you don't know when she was bred, you don't have a due date. A doe needs to stay with a buck for at least twenty-five days to be sure that she went through at least one heat cycle, and when adding a few extra days for her to deliver early or late, you have a month-long window when the doe could give birth. It is especially important to be present for births by first fresheners, does that tend to have triplets or quadruplets, and for births that are expected when temperatures are below freezing. When I was still fairly new to goats, I did pen breedings only for summer due dates.

Hand breeding isn't quite as involved as it sounds. It simply means that you put a doe with a buck when you know she is in heat. You might

Using a Buck Rag

If you don't have a buck, or if you keep your bucks far from your does, you might want to use a buck rag. What's a buck rag? It is an old piece of cloth that has been vigorously rubbed on a stinky buck so that it picks up his scent. It is then kept in a jar or sealed plastic bag to retain its scent. It has a couple of uses. First, some people say that does will not come into heat if a buck is not within sniffing distance, so they are more likely to come into heat after smelling the rag. Second, you can sometimes determine if a doe is in heat by letting her sniff the rag. If you are really having a hard time figuring out when she's in heat, you might want to let her sniff it every day. Typically, does ignore it if they're not in heat, and they want to rub on it when they are in heat.

put the two of them in a stall or pen for a couple of hours, although some people will separate them after seeing two or three successful breedings.

If you don't own your own buck, you should call the buck's owner as soon as you see any signs of heat. I've heard a lot of people say that they thought they could take their doe to get bred the day after they started seeing signs of heat because they'd read that a doe is in heat for twenty-four to forty-eight hours. Although it's true that most does are in heat that long, you don't know exactly how long she's been in heat when you notice signs. If she went into heat the previous night, she will have been in heat for twelve hours by the time you notice her in the morning, and she could be done by the following morning. Color of discharge is one clue about what stage of heat the doe is in. If it is clear, she is usually in the early stages, but if it is white, she is near the end, which means you need to hurry up and get her to a buck for breeding as soon as possible.

Even though I have my own bucks, I put a doe and a buck together as soon as I realize she is in heat, and I leave them together until the doe is no longer showing signs of heat. When you look into the breeding pen and see the two goats rubbing on each other, the doe is in heat. When she runs to the door or gate, jumps on it with her front hooves, looks you in the eye, and screams, that translates as, "Get me out of here! Do you know what that stinky buck is trying to do to me? He won't leave me alone!"

Every now and then you may have a doe that simply does not like the buck you have chosen for her. She is showing every single sign of being in heat, but when you put a buck with her, she runs from him or butts heads. If she is in standing heat and was standing for another doe or wether to mount her before you put her with the buck, she simply may not like the buck. Sometimes this happens with young bucks that are not stinky yet. And it may happen when you put an inexperienced buck that isn't quite sure which end to mount with an older doe. If you want this particular buck and doe breeding, you can leave them together for the day and hope he figures things out, provided the doe isn't being too hard on him. Some does will get violent with young bucks, and in such cases they should not be left alone together.

Buck Behavior

If you are new to goats, normal buck behavior can be shocking. They stomp and make a variety of whooping and blubbering sounds when they're near a doe in heat. It has been said that it sounds eerily human. It is common for a buck to make the Flehmen response, lifting the upper lip, when trying to figure out if a nearby doe is in heat, and a buck will sniff a doe's urine as she is peeing, sometimes even getting it on his nose. He may also pee on his front legs and face.

Successful Breeding

I've never met anyone who wasn't surprised by how quickly goats breed. It is only a slight exaggeration to say that if you blink, you will miss it. It takes only a few seconds. I once sold goats to a woman who kept complaining that they were not getting bred. Several months later she called in a panic because one was in labor. She had missed the breeding because, not realizing how quickly goats breed, she had gone into the house for fifteen minutes while the doe and buck were together.

We always wait to see at least one successful breeding before leaving a doe with a buck so that we know she is in heat and likes the boyfriend we selected for her. A successful breeding is signaled by the doe arching her back. Her front and back feet will be unnaturally close together. The buck is usually able to breed again in five to fifteen minutes. A well-nourished buck in his prime should be able to breed several does in a single day and have a sperm count sufficient to get all of them pregnant. A young buckling may not be able to successfully breed more than one doe a day unless the breedings are several hours apart. After six or seven years of age, a buck might still be able to breed several does, but his sperm count may be going down, which would lead to the does that are bred within a few hours of the first does not getting pregnant. An older buck is not ideal for pen breeding because does tend to come into heat in groups, and the older buck might not breed all of those that come into heat in one day.

Occasionally a goat will come into heat five to seven days after being bred, although it is not common. If you think you see signs of heat, don't think you're losing it. The doe should be bred again. In my experience, the due date from the second breeding is when the doe actually kids, but mark both days on the calendar just to be safe.

If the doe comes into heat three weeks or more after being bred, you should assume she is not pregnant and breed her. Again, in most cases, it is the last breeding that actually impregnates the doe, but mark the calendar for every possible due date and check for signs of impending labor as each possible due date approaches. The first time I had a doe with false heats, she continued to come into heat every month, and I kept moving her due date. I was expecting her to kid in September, and late one afternoon in July, we found her in the pasture with three kids.

If a doe doesn't come back into heat, you can assume you will see kids in five months. An ultrasound, which can be done three weeks after breeding, or a blood test, which can be done thirty days after breeding, will confirm pregnancy.

Goats usually don't look pregnant until at least three months, although I've had a few that didn't look much wider than normal the day they gave birth. Long-bodied does do an excellent job of hiding kids. On the other hand, I have a doe that's been giving birth to quadruplets or quintuplets for the last four years, and she starts to look pregnant by two months. First fresheners usually start to get udder development about a month before the due date, with the udder gradually growing as they get closer to the end of pregnancy. Senior does may appear to have a full udder as much as a week or two before they actually give birth.

Artificial Insemination

If you don't have access to a buck, you might consider artificial insemination (AI). When you consider the cost of buying a buck and feeding him, AI is less expensive. Although a semen storage tank can cost as much as a couple of bucks (male goats, not dollars), it will last a very long time. If there is another breeder in your area, or even a cattle ranch, that uses semen, you might be able to rent space in their storage tank. Semen costs about one-tenth as much as a buck. If you're lucky, you might have an AI technician or a vet in your area who can inseminate your does. The procedure can be tricky, and success is not guaranteed, but workshops teaching you how to do it yourself are available.

Feeding for Fertility

"Flushing" is the practice of feeding does more than usual for a month prior to breeding with the idea that it will increase their fertility. This is a very old practice, and the idea behind it is a good one. Of course, you want your doe to be in top condition prior to breeding. She is more likely to get pregnant and stay pregnant. However, for most people, flushing translates as more grain, which is not the best thing for a ruminant. Flushing should not be necessary in a goat that is in good condition. Feeding more grain than her body needs will simply cause her to put on more weight.

This really boils down to the fact that excellent nutrition can help a goat to perform at the highest level possible, given its genetics. But nutrition can't trump genetics and cause a goat to release more eggs than it is genetically programmed to release. For example, if you have a doe that consistently produces singles when her dam produced more, perhaps there is a problem with nutrition. But you can't expect flushing to cause a goat to produce triplets or quads when her mother and grandmothers always produced twins. The majority of dairy goats produce twins, and feeding grain prior to breeding doesn't increase that number.

I have never used flushing prior to breeding my goats, but I have seen firsthand that nutrition plays an important role in fertility and production. When we had a problem with copper deficiency, we had goats not coming into heat, not getting pregnant when bred, and not staying pregnant. However, once our copper deficiency problem was resolved, our fertility rate shot up considerably. In our Nigerian Dwarf goats, the average number of kids went from 2.5 per doe to 2.9 per doe. If I include the does that aborted or didn't get pregnant, the average was less than two kids per doe, and from an economic perspective, those does should be included because they are being fed for a year and are not producing kids or milk. Our two remaining LaManchas went from not getting pregnant at all to both having triplets.

The problem with the concept of flushing is that it puts an emphasis on nutrition during a single month. The goal needs to be optimal nutrition every month of the year. Even if I had been flushing my does during those years when we had a copper deficiency problem, it would not have eliminated the fertility problems that we were having. Although the problem was nutritional, the goats did not need more calories; they needed more copper.

The nutritional needs of bucks during breeding season definitely increase. Although most sources recommend bucks not be fed grain because of the risk of urinary calculi, you need to look at your bucks to determine whether to feed grain. For years I was so worried about urinary calculi that I didn't give my bucks grain, even though they would lose a lot of weight every year during breeding season. I talked to other breeders who lived in even colder climates than I do, and since they weren't feeding grain, I didn't think I needed to do it either. Then one day I saw one of those breeders at a goat show. Her hay was far superior to what I had been able to find at that time, which meant her bucks were getting a more nutrient-rich feed than mine, which explained why they were so much healthier than my bucks, even though none of her goats was eating grain.

Now I watch my goats. The feeding plan for my bucks changes from one year to the next depending on what the body condition looks like. If they start to lose too much weight, I give them grain. This usually correlates to the quality of the hay and how cold it gets, but the body condition also tends to be worse when they can see the does because they spend too much time fighting with each other and burning calories. If you need to feed grain to your bucks on a regular basis, add ammonium chloride to their grain or minerals to prevent urinary calculi or feed a brand of goat feed or minerals that already includes ammonium chloride.

CHAPTER 9

Pregnancy

✦ ✦ ✦

B REEDING A DOE is just the first step towards greater self-sufficiency, but there is a lot that needs to be done between breeding and that first batch of chèvre.

Gestation

The first thing to do after breeding your goat is to mark your calendar with her due date. Goats are normally pregnant for five months, but I mark my calendar for 145 days because they can give birth to perfectly healthy kids a few days earlier, and I don't want to get surprised. Standard goats can give birth as late as 155 days, but the majority of Nigerians give birth by 150 days.

Signs of Pregnancy

Although it is widely accepted that you should stop milking a doe two months before her due date, you will usually see a reduction in milk production by the time a doe is two months pregnant, and she will have mostly dried off by three months. Like everything else with goats, however, this is not absolute. A breeder who had a doe jump off the milk stand one morning, lie down in the barn and push out a couple of kids now does blood tests to determine whether the does are pregnant.

Many people wonder when their doe will begin to look pregnant. In most cases, the abdomen won't look bigger until she is at least three months pregnant, although a few long-bodied does are especially good at hiding kids and may keep you guessing until you see an udder starting to form. My Nigerian doe Coco started to look pregnant by two months both times she was pregnant with quintuplets.

False Pregnancy

It is possible for a doe to get bred and stop cycling and even to get a big belly and develop an udder and appear to be pregnant in every way, yet

When most people see a doe with a stomach like this, they assume she is pregnant, although she is not. This is a dropped abdomen and is not a sign of illness. Basically, after having kids for a few years, she has lost her girlish figure. Her abdomen got stretched out one time too many, and it will never regain its original shape.

The same doe is pregnant in this photo. The difference between a pregnant abdomen and one that has simply lost its shape is that the pregnant one is much higher and fuller.

not be pregnant. "False pregnancy" is used synonymously with "hydro-metra," which simply means water in the uterus. Because the hormones are involved, a blood test shows a false positive. An ultrasound examination is the only foolproof way of determining pregnancy, but blood tests are still popular because they are less expensive and breeders can learn to draw blood themselves, reducing costs further. Because false pregnancy is rare, blood tests are still considered very reliable. A false pregnancy may not last for five months. It usually ends in a "cloud burst," which is basically a release of all the uterine fluids without a kid or placenta.

Some false pregnancies started with a real pregnancy that terminated very early but the body didn't recognize there was no longer a fetus. However, in some cases of false pregnancy, the doe has not even been exposed to a buck.

Feeding During Pregnancy

Does should be neither underweight nor overweight when they are bred. Pregnancy is not the time to put a goat on a diet or try to put weight on her. Assuming that your does are in ideal condition when they get pregnant, you don't need to make any immediate adjustments to their feeding program. You should continue feeding milkers as usual until they are dried off at three months gestation.

During the two-month dry period, the doe should be fed a total dry-weight ration that is equal to about 2 to 2.5 percent of her body weight, and no more than about 0.5 percent should be grain. Feeding too much grain in pregnancy can result in overly large kids, especially if it is a single fetus. If you have good quality hay, you may not need grain until the last couple weeks of pregnancy when you start feeding it to acclimate the goat to digesting grain again. Typically, the goal is to be feeding a doe half of her milking ration by the time she kids. Always make changes to a goat's diet gradually. If you suddenly increase grain intake for a doe that just kidded, odds are good that she will get diarrhea, even if she doesn't wind up with one of the more serious digestive disorders.

It is said that you should not feed alfalfa to pregnant does, but this is based on information regarding dairy cattle. For more information on this, see "Hypocalcemia" in Chapter 7.

CHAPTER 10

Birthing

✦ ✦ ✦

GOAT OWNERS worry about kidding, even though it goes well almost every time. One study found that intervention is needed in only 5 percent of births. If you find yourself "helping" too often, it means one or more things: you are overreacting, you need to re-evaluate your nutritional program, or you need to cull the whole herd and start over with different genetics.

Intervening always carries a risk of infection or damaging the doe's uterus or even of hurting the kid, so it should not be done routinely. Regularly intervening also means that it will become impossible to selectively breed goats for birthing ease. You will have no idea if a doe can give birth on her own if you aren't letting her do it. Most of my does never need assistance. I'm just there to get the kids dried off so that they don't get hypothermia, and then I make sure that I see them nurse before I leave the barn.

Unfortunately, things do go wrong sometimes, but you won't know that unless you know what is normal and know when you need to intervene or ask for help. It is a good idea to have the phone number in the barn or in your cell phone for a vet that has experience with goats. Hopefully you also have a mentor, an experienced breeder who is willing to answer your questions as they come up. As you read through this section, try to focus on the fact that most births are completely normal and you'll need to do little other than dry off the kids and coo over them.

Getting Ready

Like so many aspects of raising goats, if you talk to ten people about their kidding set-up, you will get ten different answers. We started with only three does, and we left them together during kidding because I had read that other does can help clean off the kids if a doe gives birth when you aren't there. This can be especially helpful if you have a first freshener or a doe that has triplets or quadruplets, which can be born with so little time in between that the doe may not be able to get their noses cleaned off quickly enough to avoid suffocation. However, if you have any questions about the disease status of any of your goats, having does together for kidding is a bad idea because a number of diseases can be passed through birth fluids.

After a few years of kidding in an open barn, and realizing that not all does are totally supportive of other does in labor—and some are downright mean—I decided I wanted to have kidding pens. In addition to does being more protected, kids born in a pen have an easier time adapting to life outside the womb. A kid is born with an instinct to nurse,

Our semi-private kidding suites use pig panels to separate the pens. Because goats are herd animals, they get stressed when they can't see other goats, so by using pig panels, the does are able to see each other, but they are also protected from being picked on. Two does will sometimes butt heads through the panels, but one cannot slam into a pregnant goat's belly. This doe gave birth a couple of hours ago and is now passing her placenta.

but it has nothing telling it that it's not okay to nurse from a doe that is not its mama. It is a rare doe that will let a kid nurse if it is not hers. It is not uncommon for them to butt or bite a strange kid that tries to help itself to a snack. Kidding pens are smaller than a big barn, so it is impossible for kids to get lost. It is also easier for me when a doe gives birth in a smaller space. Some does walk when they're pushing, and more than once, I found myself following a doe on my hands and knees with a towel as a kid was emerging. My favorite thing about kidding pens, though, is that does are giving birth in a clean space because I don't move them in there until they are within a day of giving birth.

Vicki McGaugh has been raising goats for twenty-seven years, and during part of that time, she was operating a dairy that included some goats that were CAE-positive. For that reason, she used kidding pens. However, after more than two decades of raising kids on pasteurized milk and heat-treated colostrum, she now lets her does kid in the main barn, although she still attends all births and bottle-feeds the kids. She prefers letting does give birth with the other goats around because she feels it's less stressful for them to remain with the herd. "Until the does were actually delivering, my does hated to be away from the herd," Vicki says. "It also made for some tussling around as a doe that had spent several days in a kidding pen moved back into the herd."

Although everyone has their own preference when it comes to using kidding pens or letting their does kid in an open barn, the one thing that is absolutely necessary is that the area where she gives birth needs to be clean. If you are using the deep bedding system, you need to be adding a new layer of straw every day or two. Remember that kids can aspirate or choke on shavings. There shouldn't be exposed goat berries, and the exposed straw should not feel wet or appear to be saturated with urine. When using deep bedding, the barn needs plenty of fresh air without being drafty.

Kidding supplies

You probably already have many of the following items in your home or barn. The more specialized supplies can be bought through the various goat and livestock supply companies online.

- **Towels:** This is listed first because it is the single most important thing to have at every birth, especially if it is cold outside. You might want to use old towels, or you can find fairly inexpensive bath towels at discount stores. Some people use newspaper to remove the initial birth goo from kids before toweling them off.
- **Baby monitor:** Unless you live in your barn, you need a baby monitor so that you can hear what's happening out there when does are due. The least expensive baby monitor will probably work. We have a metal barn, and the signal doesn't go through metal, but we put the transmitter near the door, and it works well. A screaming goat echoes throughout the barn, so the monitor picks it up 60 feet away, and we hear it clearly in the house.
- **Blow dryer:** Somewhere around 10°F–15°F, a blow dryer becomes necessary to get the kids dried off so they don't get hypothermia.

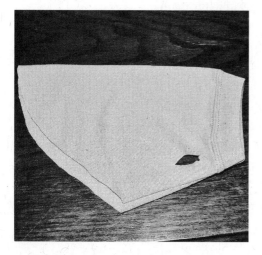

Measure a kid from the back of its head to its tail. That's how long the kid coat needs to be. The wrist of the sweatshirt arm is the collar of the coat. Cut two small holes, one on each side of the seam, for the kid's legs. When putting it on a buckling, be sure the bottom of the coat does not cover his penis, or he'll soon be wearing a wet coat.

Ears and tails need special attention because they can freeze if they are wet. If the temperature is below zero, a kid's wet ears can freeze within a few minutes.

- **Heat lamp:** In most parts of North America if it's below freezing, you will need a heat lamp for only the first couple days after birth. After that kids are able to maintain their body temperature until temperatures get close to zero Fahrenheit. If you use one, be sure it is securely attached to a wall or permanent structure so that it cannot be accidentally knocked down. Heat lamps are the number one cause of barn fires. Do not use a heat lamp unless it is truly needed.
- **Kid coats:** If you don't have electricity in your barn and temperatures are getting close to zero Fahrenheit, you might want to make little coats for your newborn kids. Be sure the kids are completely dry before you put the coats on them. They can be made from old infant T-shirts for standard goat kids and from the arm of adult sweatshirt for Nigerian Dwarf kids.
- **Bulb syringe:** If you have had a baby, you might have one of these already. You don't need it for most goat births, but if you hear a lot of gurgling when a kid starts to breathe, you might want to use a bulb syringe to clear the nose. Always remember to depress the bulb BEFORE you put the tip of the syringe in the nose so that you don't accidentally force mucus into the lungs.
- **Iodine:** The end of the umbilical cord can be dipped in iodine to prevent infection. An old-fashioned film canister or a small prescription pill bottle works well for holding the iodine for dipping. Iodine is also used to disinfect fingers or a hand that will be put into a doe's vagina, even if you are wearing a disposable glove.
- **Scissors:** Although an umbilical cord usually will break spontaneously as the kid is being born, it doesn't always. Some people will cut the cord with scissors, but my practice is to tear the cord a few inches from the body, leaving at least 3 inches of it attached to the kid.
- **Disposable gloves:** You don't have to wear gloves for an average birth unless you want to protect yourself from a doe's birthing fluids. If you have any doubts about her disease status, or if you are pregnant, you should wear gloves. You should also wear gloves if you need to

put your hand into a doe because it is virtually impossible to get your bare hands very clean.

- **Pritchard teat and bottle:** It's a good idea to have these on hand just in case, especially if they are not available for purchase locally. They cost only a few dollars, and if you raise goats for long enough, you will need them. Our first kidding season went perfectly, but we wound up with two bottle babies our second year. In a pinch you can use a human baby bottle, but you'll need to cut an X in the nipple to create a hole big enough for a goat kid.
- **Feeding tube and 60 cc syringe:** You won't know you need one of these until the minute you need it, so have one on hand. I've gone as long as four or five years without needing to tube feed a kid, but I've also had years when I had to tube feed two or three.

Signs of Labor

Figuring out when a doe is going to kid is probably the most challenging thing to learn when it comes to raising goats. I've seen so many new goat owners think their doe is in labor because she's breathing hard or standing funny or even because of the way she's looking at the person. I know from experience that it is very easy to project our own feelings onto a doe—those hopeful feelings of seeing new kids soon! The good news is that you don't have to be a goat whisperer to figure out when your goat is going into labor. There are several signs that she is getting close to kidding, but keep in mind that every goat is different, and the same doe may give birth differently from one year to the next.

Udder

A doe will start to get an udder during the last month of her pregnancy. It gradually gets bigger, and then, typically, it suddenly looks full a few hours before she kids. There have been times when I was doing chores and a doe turned to walk away, and her udder jumped out at me, making me say, "Whoa!" A big change that happens very suddenly usually means the doe is getting close to kidding. Some does, however, will get a very big udder as early as day 142. For these does, the udder may not be the best indicator of imminent birth.

Tail ligaments

Tail ligaments are often mentioned as a way to figure out if a doe is close to kidding. It can take years to learn to read tail ligaments accurately, so don't get discouraged if you don't "get it" quickly. During our third kidding season, we thought a doe had lost her ligaments two weeks before her due date. We asked online if this was possible and quickly got a couple of responses saying yes. But after a few more years of experience, we realized the tail ligaments don't soften up completely until a doe is within about twelve hours of kidding. We simply had been having a hard time finding them as the muscles around the ligaments had softened.

The ligaments at the tail head are normally so hard that they can easily be mistaken for bones. However, as a doe gets closer to kidding, the ligaments will start to soften up. A lot of people say they "disappear," but that's not entirely accurate. As they get softer, the ligaments get harder to find, but of course, they're still there. Confusion sets in because the muscles in that area also get softer, such that it is sometimes possible to wrap your fingers around the tail head several days before a doe kids. But this doesn't mean the ligaments are necessarily softened, only that the muscles have softened.

If you check a doe's tail ligaments a couple of weeks before she is due to kid, you will know exactly where they are and be more likely to notice the subtle changes that occur. It is helpful to compare does if you have several due around the same time. Learning to read tail ligaments is easier if you have someone to compare notes with. My two daughters and I frequently disagreed on the softness of a doe's ligaments. One of us would pronounce a doe's ligaments "gone," but then another of us would go out to the barn and come back saying that she had found them right away.

I often see comments online that a doe's ligaments were gone but returned the next day. Because the softness of the ligaments is connected to the doe's hormone levels, this would be cause for concern if it had indeed happened. However, in every case I've seen, the doe gave birth to healthy kids several days later, leading to the conclusion that the commentator simply had missed the presence of the tail ligaments a few days earlier.

Belly

One of the things we started to notice over the years was that when a doe was within hours of kidding, her belly would "drop." During most of a pregnancy, a doe's belly is very high. My youngest daughter would describe a doe that was not looking like she would kid soon as "still a table top." As the kids start to jockey for position and line up to be born, they fall into the lower part of the abdomen, the upper part of the abdomen looks less round, and there is a hollow appearance near the spine.

The appearance of the belly and the udder, and the condition of the tail ligaments, give a reasonable warning that the doe will kid within the next twelve to twenty-four hours. Once you see all three of those changes, you will start to see other changes that mean the doe is in early labor. The doe will start to act differently when she is in labor, but you cannot assume she is in labor based on personality changes alone. So many new goat owners assume a doe is in labor based solely on her behavior, and I include myself in that group. The first couple of years that I had goats, I spent many hours out in the barn sure that a doe was going to give birth soon because of the way she was acting. If you have a doe in a stall or pen by herself, odds are high that she will start to bleat when you leave her, regardless of whether she is anywhere close to giving birth.

Personality changes may include a doe that is normally aloof suddenly wanting your attention or the herd queen that is normally pushing other goats around now ignoring the herd members. As a doe gets closer to giving birth, she will lose interest in food. Goats normally love grain, so if you offer grain to a doe and she completely ignores it, she is probably within an hour or two of giving birth.

A tiny bit of mucus on her vagina doesn't usually mean much, but a string of mucus an inch thick and several inches long usually means a doe is within minutes of kidding.

Mucus

As the doe gets closer to kidding, you may or may not see a string of mucus hanging from her vagina. Don't get too excited if you see a little bit of mucus glistening on a doe's vulva. On the other hand, if a doe is scream-ing loudly but you see no mucus, don't assume that nothing is happening. Mucus is one of the least reliable indicators of labor.

Birth

When a doe begins actively pushing, you should have no doubt about what is happening. Does can be quiet and stoic in the early stages of pushing, or they can be talkative early in labor, but at some point, al-most every doe will start to vocalize quite loudly. Although most does will lie down when they start seriously pushing, they may stand up and lie down again to reposition themselves over and over again, and some give birth standing.

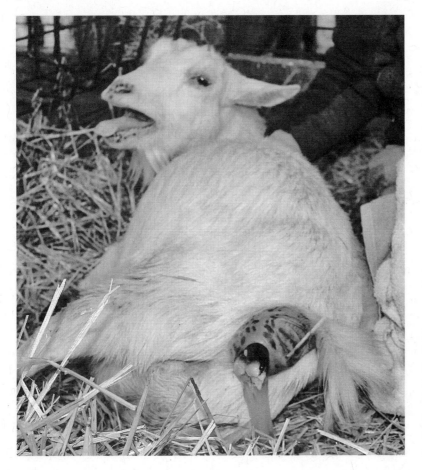

Although some does give birth standing up, if she is lying on her side, she will be pushing her legs straight out in front of her body. Her ears may be back, and she may be vocalizing loudly.

Pushing time

How long do goats push? The answer to this question can be incredibly varied! It can appear that some goats push for less than five minutes and a kid torpedoes out of its dam's body. Those are usually fun for new goat owners because there is no time for worry. Some goats may be screaming for fifteen minutes or more before you see a nose or a hoof, however, and it can be challenging to relax. The important thing to remember is that there is no reason to rush. Seconds rarely make a life-and-death dif-

---

What I learned from Fannie, Gherri, and Star

In the weeks leading up to our first goat births, I was reading everything I could find on the subject. My two youngest children and I were sitting in the barn with Fannie when she was in labor. I was holding one of my books, continuing to reread sections and looking at drawings that showed the different ways goat babies might try to be born. The book implied that anything other than two front hooves and a nose would be a problem that would require our assistance.

As Fannie pushed, we finally saw a nose and then a hoof. And as she continued to push, we still saw only a nose and one hoof. Based on what I had read, I thought that one of us would need to put a hand inside Fannie and pull the second front leg forward. My then ten-year-old daughter and I began to freak out. "Are your hands clean?" I asked her. We quickly discussed who should go wash their hands to help Fannie, but as we were talking, the goat pushed out the kid.

A couple of days later, we walked out to the barn in the morning to find Gherri had given birth to two healthy kids that were bouncing around and nursing. When our third doe went into labor, we were again there to watch and help out, if necessary. After the first kid was born and we dried it off, the second kid presented tail first. As we began to freak out again because the kid was not presenting "correctly," Star gave a big push, and the kid came shooting out before we could do anything. I started to wonder if perhaps the books had exaggerated the need for assistance and started to relax about birthing.

ference. Before seeing part of a kid, it is hard to know when you should start to worry because of the difference in the way that goats act during birth. If a doe is screaming loudly during contractions every few minutes and laying her head in the straw between contractions, you may start to think about intervening after half an hour. But if a doe is quiet and appears to give a wimpy little push every fifteen minutes, she might be fine for a few hours. You know a birth is going well when you see a hoof, nose, or tail, and progress with each contraction. Ultimately, experience is the best teacher.

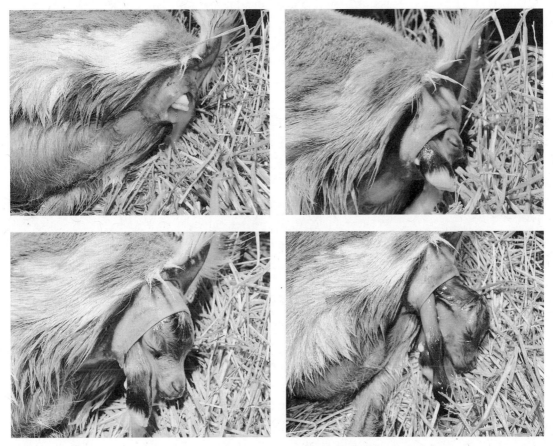

This is a textbook presentation of a kid at birth. First you see the front hooves, then the nose. In this particular birth, five minutes elapsed from the time the first photo was taken until the time of the last photo, and the entire kid was born less than a minute after the head emerged. Some births will go more quickly, however, and some more slowly.

Amniotic sac bubble

You may or may not see a bubble of fluid emerge. There is no need to pop it as it will usually break on its own, so you don't want to be terribly close to the doe when a bubble is emerging or you could wind up getting wet. Many times as the doe pushes out a kid, it may be in an amniotic sac. If the sac doesn't break as the kid is being born, you need to break it and clean off the kid's nose so it can take its first breaths. This is one of those simple things that can lead to a kid's death if you aren't there during birth. Although experienced does are usually very good about cleaning up kids at birth, a first time doe may not realize what is happening and won't know that she needs to turn around to start cleaning the kid. If a doe has multiple kids, the subsequent kids may be born so quickly that the doe hasn't cleaned off the first one by the time the others are born. This is more likely to happen with triplets or more.

Malpresentations

Determining when you need to assist at a birth is not an exact science, and you can never go back and redo something that did not turn out well. That means that once you do intervene, you'll never know if the situation would have resolved itself without your assistance. There is no book that can tell you what to do 100 percent of the time when you are with a doe that is giving birth because every situation is unique. Sometimes even the most experienced goat owner is challenged to know what the best course of action is for a doe that appears to be having difficulty. And, of course, nothing can replace the services of a good veterinarian.

A seemingly lengthy labor without any sign of a kid is commonly a worrisome situation, and sometimes the reality is that the owner has assumed the doe was actually in labor when she was not. I have even known several people who sat up all night with does they thought were in labor when it turned out that they were not even pregnant. It is not unusual for new goat owners to think a doe is in labor for days before she actually gives birth.

The most common birthing complication is a misaligned kid, and fixing it is more art than science. A kid needs to have one end of its body presenting, so if you can see a hoof, a nose, or a tail, the situation is not too bad. With each push, you should see a little more of the kid's body part. If a doe has gone through several contractions with no progress, it may be a sign that something isn't quite right.

There is no shortage of forums and Internet groups where you can sign on and start asking questions or posting comments about goat keeping. This can be a wonderful way to learn and share information, but there is also the possibility that you will receive some seriously bad information. Answers are given to questions about situations that may not be clearly described. In the case of birthing complications, a lot of people are quick to suggest intervening in a birth without getting a clear picture of what is happening. As a new goat owner, I was on the receiving end of some bad advice a few times. I signed on and said that I had a doe that had been in labor for a couple of days, and several responses painted terrible pictures of what horrible tragedy must have already occurred. Some strongly encouraged me to intervene and start pulling kids. Had I tried, I would have been unsuccessful because the reality was that the doe was not even in labor. She gave birth to three healthy kids the next day.

If you use an online forum for goat advice, always include as much information about a situation as possible. I've seen people ask something like, "How soon after a doe passes mucus will she give birth?" which seems like a simple question. People may immediately start to respond, but the answer to that question is that it could be anywhere from a week to five minutes depending on how much mucus was actually passed.

When posting a question online, don't simply say that a doe is in labor. Describe exactly what she has been doing and for how long. Is she eating and drinking? Vocalizing? Standing? Walking? Lying down, looking alert, with her head up? Or is she lying with her head on the straw and panting? Even if you include as many details as you can think of, the people giving you advice may still not fully understand your situation. And remember that you have no way of knowing how much experience and knowledge anyone online has.

Getting Help Online

Noses

A lot of people worry if they don't see a nose and two hooves. Maybe this is a difference in breeding stock, but in my experience does have no trouble giving birth with a nose and one hoof or a nose only unless the kid is very large. They may need to push a little longer than if the kid had two hooves and a nose presenting, but the kid will usually be unaffected. We learned early in our homesteading days that baby animals seem to weather the most tumultuous births with less stress than the humans who are watching. We had a rather wild yearling ewe running around the pasture for forty-five minutes with a lamb's head sticking out of her back end as we tried to catch her. The lamb was just fine when we finally got it out.

If only a nose is presenting and a doe is not making progress, or if she births the head and then makes no progress after several contractions, you may need to reach in, find one of the legs, and pull it forward to decrease the circumference of the kid's shoulder area. If the kid's head is already out, you may have to push it partially back in to have enough room to find the kid's front leg. This can be physically challenging because it can be a tight squeeze to get your hand into the cervix. And you have to be very careful because a doe's cervix is paper thin when fully dilated, and it is easy to tear it. The most important thing to remember in this situation is to not panic. There is no reason to hurry.

If the nose and one front leg are out, you may be able to simply put gentle traction on the leg as the doe is pushing.

Breech

If you see a tail presenting, it will take longer for the doe to push out the kid because the back end of the kid is blunt and larger than the nose and two front feet. Breech births are not uncommon with goats, and they are not usually a cause for alarm. Most goats are able to give birth to a breech kid without assistance. If you see a tail presenting and the kid is not making any progress after several contractions, you can reach into the doe to pull the hind legs out, reducing the circumference of the kid. The biggest danger with a breech birth is that the kid can wind up deprived of oxygen if the head is not born in a timely manner after the umbilical cord is out. Once the back half of a breech kid has been born, the placenta will not be supplying oxygen to the kid because the umbilical cord is pinched or broken. You need to gently pull the rest of the kid out of the doe's body so that the kid can start breathing. If the doe is standing, gravity will usually take care of this if the kid is heavy enough.

Hooves

Ideally, hooves should be coming out of the doe's vagina in such a position that if the doe were standing, the kid would be exiting the vagina and stepping onto the ground (if it could walk the moment it's born, of course). Upside-down hooves are most likely hind legs, and the doe will be able to give birth more quickly than if the kid were tail first. Back hooves presenting is the most stretched-out, tapered position possible for a kid being born.

The front hooves can also present upside-down. If you see upside-down hooves and the doe is not making progress, it could be the kid is upside down, or posterior. To figure out if you are seeing front or rear hooves, run a couple of fingers up along the legs. Front legs will have knees where the legs bend towards the back of the hooves, whereas hind legs will have hocks that bend in the direction of the front of the hooves. If you discover that a kid is posterior, you should call your veterinarian or mentor.

Ribs

When it comes to birthing, anything is possible, but unlikely. More than two hundred kids had been born on our farm before we had our first truly unusual presentation when a kid tried to come into the world ribs first. As you can imagine, that didn't work. I had to turn the kid so that the nose and a front leg were presenting, which was more stressful mentally than it was physically challenging. Since then we've had a kid born with the crown of its head first, rather than its nose. The kid was much smaller than average, which was good because the crown of a kid's head is usually too large for a doe to push out.

Head

A presentation with two front feet sticking out and the head turned around over the back is problematic. This presentation is a perfect example of why you should never pull on a kid unless you know where all the parts are. If you pull on the front legs of a kid whose head is turned backwards over its spine, you will make no progress, and you may wind up injuring the doe and the kid. To correct this situation, the head needs to be turned so that the nose is pointing forward.

Two kids

Two kids presenting simultaneously is rare and sounds worse than it is because at some point one kid will finally slip past the other one and be born. However, in the interest of not exhausting the doe, you can hold back one kid as the doe continues to push. Hopefully, you have two kids trying to come out front hooves and noses first. If you have one breech and one nose and hoof first, it is a good idea to help the nose and hoof come out first. If you pull the breech kid first, its head could lock with the head of the other kid and get stuck, which would necessitate pushing it partially back inside so that you can hold back the head of the second kid. In the amount of time it takes to do that, the breech kid could die from lack of oxygen as the umbilical cord would have been pinched or torn.

Failure to Dilate

It is rare that a goat will fail to dilate, but it can happen. The only caesarean section that we've had so far was in a six-year-old that didn't dilate. This is sometimes called "ring womb," and in less severe cases, you may be able to manually dilate the cervix. However, you should talk to an experienced goat vet before doing this, and keep in mind that it is possible to rupture the uterus or tear the cervix.

Umbilical Cord

When a kid is being born, the umbilical cord almost always breaks a few inches away from the kid's body. If it doesn't break, many people cut it, and in either case, will tie off the cord with string or dental floss, and dip the end in iodine. I followed this procedure for the first few years I had goats, and I even used fancy plastic cord clamps instead of dental floss or thread. Then one day I witnessed the most interesting argument on a goat Internet group about how to treat umbilical cords at birth. On one hand, people argued that kids would get navel ill and possibly die if you didn't dip the cords. On the other hand were people who said it was a waste of time because the does would lick off the iodine. Some people were dipping cords multiple times over the first day or two, while others had completely given up the practice. The interesting thing is that none of those people who were not dipping the cords had ever had a kid with navel ill. I started looking for research on dipping cords and found a sheep study that showed a lower rate of infection among lambs that were pasture born than those that were born in a barn and had their umbilical cords cut and dipped.

Assisting in a Birth

Before putting your hand into a doe's vagina, clean off her back end by washing it with warm soapy water or by using gauze pads to clean it with iodine. Always wear gloves. If you have Nigerians, you can probably use disposable gloves available in any drug store, but if you have standard goats, you will need cattle OB gloves, which cover the arm up to the shoulder. I always squirt iodine all over my gloved hand before putting it into the goat. If you need to insert more than a couple of fingers, you should use a lubricant, which can be a natural oil, such as olive or sunflower. Obstetrical lubricants are also available, but I prefer not to use them because they're a blend of various chemicals.

Ultimately I developed my own middle-of-the-road approach to umbilical cords. The lamb study basically proved to me that even if you dipped cords, it would not compensate for a dirty environment at birth, which is one reason I wanted to have kidding pens for my does. It also occurred to me that if you were going to cut the cord with a pair of scissors, they needed to be sterile, and the cord needed to be sterile so that you were not driving bacteria into the cord as you cut it. It is not easy to keep a pair of scissors sterile when you're in a barn, nor is it easy to sterilize them between kids. In my research I had read a suggestion that tearing the cord off a few inches from the belly rather than cutting it short, mimicked nature, so I adopted that as my practice. On the rare occasion a doe surprises me and gives birth somewhere that is not clean, I dip the umbilical cord in iodine, and in the case of an umbilical cord that tears off at the belly during birth, I douse the belly button with iodine.

Regardless of whether you have dipped the cord, it should be left alone to dry up and fall off on its own over the next week or two.

Placenta

In most cases of multiple births, there is nothing hanging out of the doe's vagina between kids, but after the last kid is born, there is an umbilical cord and other birth membranes. Within three or four hours after the last kid is born, the doe will pass a placenta. There will only be one placenta, regardless of how many kids were born. Although some people take it away immediately, I let my does eat it if they are interested because it is rich in nutrients. If they have not eaten it within an hour or two, I remove it from the birthing area. Some caution that a doe can choke on the placenta, and although it is possible, it is very uncommon.

After the doe gives birth to the last kid, you will see an umbilical cord or membranes from the amniotic sac hanging out of the doe's vagina. Over the next few hours, you will start to see the placenta as each of the cotyledons separate from the uterus.

The placenta can take as long as twelve hours to pass, but that is unusual. Do not pull on the placenta or attempt to rush it by tying something heavy on it. If you tear it, it could break off inside the vagina, and you will never know whether the doe has passed the remainder until you have a very sick goat. If a doe has a lot of membranes hanging out, she may try to grab them and pull on them herself. If one of my does tries that, I tie the membranes in a knot (or two or three) to make them short

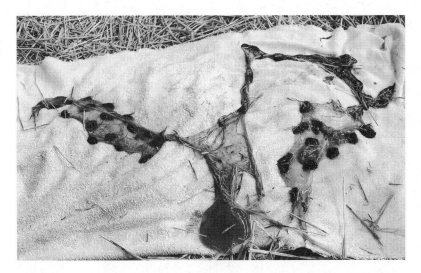

A goat placenta looks like a lot of little red prunes connected to each other by a clear to white membrane. The prune-looking things are cotyledons, and the nutrient exchange that occurs between mom and babies happens in the cotyledons. There are about 70 to 125 cotyledons in a goat placenta. The reason it normally takes a few hours for the placenta to come out—and why it seems to come out inch by inch over the hours—is because each one of the cotyledons has to separate from the uterus.

enough that she can't reach them. I want to know when the placenta is passed, and I won't know if the cord and other membranes are ripped off. When a doe gives birth at midnight, you might not want to stay with her for hours until she passes the placenta. If there was a cord hanging out of her when you went to bed, you will know she has passed it and eaten it during the night if you don't see any sign of it in the morning.

Feeding Post Birth

After the doe gives birth, she is usually thirsty and hungry. If it is cold outside, I get a bucket of warm water, which most does love to drink in the middle of winter. I also give her a pan of grain. While the doe is eating her grain, she is standing still, so it gives me the opportunity to work with any kid that has not yet perfected its nursing technique. If a doe had multiples, I will also put the smallest kid under the doe to give it another opportunity to nurse. If you will be bottle-feeding the kids, put the doe on the milk stand to give her the grain, and milk out all of the colostrum.

Newborn Check

Shortly after each kid is born, you should do an initial newborn exam to make sure that each kid has all its pieces in the right places. In addition to checking for obvious things, such as an anus, you also want to know

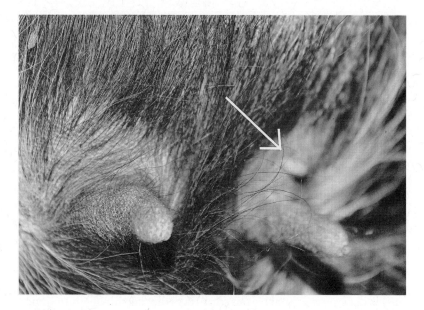

The extra teat on this young doe is growing near the base of another teat.

if a kid has any disqualifying defects so that you don't offer it for sale or get your hopes up about its future in your herd. Peeing or pooping is generally a good sign that the newborn's plumbing is in working order. Although it is rare, kids are occasionally born without an anus, and obviously they will not survive. If a kid latches on and nurses well, the mouth is probably in good shape. However, if milk comes out the kid's nose or if it has difficulty latching on, run your finger along the roof of the kid's mouth to be sure it doesn't have a cleft palate.

Check that each kid—buck or doe—has only two teats. Extra teats are a disqualification in show goats, and they are not something you want in milkers. In addition to possibly getting in the way when milking, extra teats can also get infected if they are functional. If they are not functional, kids can get confused and try to suck on them and then not grow properly because they are not getting enough to eat. A buck with extra teats should not be used for breeding, so you should plan to castrate it.

You also want to be sure that bucklings have two testicles, regardless of whether you plan to keep them intact. If testicles are not descended at birth, the odds are good that they will not descend. A buck with only one testicle should not be bred, and it is impossible to easily castrate a buck with an undescended testicle, making it a challenge to sell as a pet.

With an undescended testicle, he will still get stinky and act bucky. Most people will use a cryptorchid as a meat animal.

Kid Complications

You might think that as soon as the kids are born, you are home free, and that is usually the case. However, every now and then, you will have challenges with kids.

Weak or non-responsive kids

Most kids are born with plenty of energy. They may be screaming within a minute of taking the first breath, and some are even trying to stand a few minutes later. But what do you do if a kid seems to be barely alive? Some people recommend swinging a kid upside down to clear its lungs, but I prefer a more gentle approach. I hold the kid in my lap with its head lower than its body, and I use a bulb syringe to suction both nostrils, as well as the back of the mouth. I rub the kid vigorously with a towel, placing one hand on each side of the chest. If you decide to swing a kid, keep in mind that newborn kids are very wet and slippery, so it may be helpful to hold the kid with a towel to get a better grip. The head is held in one hand, while the other hand holds the kid's body.

Hypothermic kids

If you live in a northern climate and kid during the colder months, you will have to deal with hypothermia at some point. If you find a kid that is still wet, but is cold and non-responsive, don't be too quick to declare it dead. If you can feel a heartbeat, it isn't too late to try to save it. The goal is simply to raise the body temperature back to normal as quickly as possible. The best way to do this is to put it in a bucket or sink of warm water that is 100°F–105°F. There is no need to find a thermometer. If it feels like bath water, it's probably fine. Put the kid's body in the water with only its head sticking out. You can check the kid's temperature by putting your finger in its mouth. It will feel ice cold, but as the kid warms up, its mouth will feel warmer. Once the mouth feels warm and the kid is more responsive, you can dry it with a towel and move it to a heating pad.

Without a thermometer, it is easy to overheat a kid when wrapping it in a heating pad. Overheating will cause a seizure that the kid may or may not survive. A thermometer placed between the kid's body and the heating pad should not read above 102°F because you don't want the kid's body to heat up more than that. If you get a reading above 102°F, do not wrap the kid in the heating pad, but lay it on top of the pad. The pad will continue to heat the kid's body, but the kid will be able to cool itself by releasing its body heat into the room. Continue to check the temperature of the kid's mouth over the next few hours to make sure it is maintaining a normal body temperature, especially after you turn off the heating pad.

Most kids with hypothermia become bottle babies because the stress sets them back developmentally. After it has suffered from hypothermia a kid might not be walking around for a day or two, making it impossible for it to nurse. If you feel the kid sucking on your finger when you check its temperature, it should be able to take a bottle, but if it is not sucking, you will have to tube feed it. It is important to get colostrum into these kids as quickly as possible. It is ideal if someone else can be milking the dam or thawing frozen colostrum while you are warming up the kid.

CHAPTER 11

Raising Kids

T<small>HE DECISION</small> to dam raise or to bottle raise kids needs to be made before the kids are born. There has been a bias towards bottle raising baby dairy animals for the past few decades as factory farms took over the dairy industry. When you have thousands of cows in a dairy, it is impossible to socialize them if they are dam raised. Dairies want as much milk as possible from the cows, which they can do if they milk them and ration the milk given to the calves, selling the excess milk and increasing profits. In spite of the fact that our goat website includes milk records and clearly states that our kids are dam raised, I still get questions every year such as "Can you milk a doe if she was dam raised?" I always point out that people were milking cows, goats, and sheep for thousands of years before bottles were invented. An old-time farmer will laugh at the idea of bottle raising kids. It is a modern misconception that dam-raised kids are inevitably wild.

Like most animals, goats will be wild without plenty of human contact as babies. A litter of kittens found in the woods or in a secluded part of a barn will be as wild as lions. Goats are very much the same way. When kids are born on pasture and get little human interaction, they will be wild and difficult to handle. When handled daily, though, they will be friendly. But regardless of whether their mother or a human raises them, some kids can be incredibly stubborn.

Some claim that it is easier to milk does that were raised on a bottle. This view comes from the fact that when an entire herd is employing bottle-feeding, the does are easier to milk overall. It is not the doe that was bottle-fed as a kid that is easier to milk, but rather it is the doe that has no kids to feed that is easier to milk. If a doe has been nursing kids for a couple of months and is put on the milk stand, she may not be thrilled with the idea of letting you milk her because she firmly believes that the milk is for her kids. Just as she would kick at any strange kid that tries to nurse, she may kick at the bucket or your hand.

What I learned from Cicada

The first year that we had more than a dozen goats kidding, my daughters decided that milking ten does was more than enough. As several more kidded, we put them and their kids in the pasture across the creek. Although they had access to wonderful browse, it meant that they had human interaction only twice a day, when we refilled the water. Up until that point, we had never had any unfriendly kids, so we didn't realize the importance of spending time with them.

We kept three of the doelings, and although one of them was quite friendly, the other two were not. In fact, Cicada was so wild that it was impossible for anyone to catch her without the help of other people. By the time she was a year old, I realized it would have been impossible to milk her if we had bred her to kid as a yearling. I put her in a 10-foot by 10-foot stall in the barn with another goat, and although they had as much hay as they wanted to eat from a hay feeder, I went in twice a day with a small pan of grain, which I held in my lap. As dry does, they didn't need grain, but goats love it and can usually be bribed with it. The other doe would run up immediately and start to eat. Being competitive and curious, Cicada hesitantly inched towards me to get the grain. I let her have a few bites before I touched her. She bolted. Each day, when I touched her, she reacted a little less quickly, and by the end of the week, she decided that she could eat the grain while I was petting her. Today she is one of our best milk goats.

Aside from the issue of disease transmission, the decision to dam raise or bottle-feed kids is one of personal preference. While some people say that bottle-fed kids are friendlier, I view them as annoying. They don't realize that you are not the same species, so they are very pushy, jump on people, pull long hair, and suck on fingers, which can be dangerous if you have small children. It is very easy for a kid to suck a finger to the back of its mouth and bite, drawing blood. They are also much harder to keep fenced in because they don't have the desire to stay with mom and the rest of the herd.

In our third year of raising goats, we wound up with five bottle babies. We had been using single-strand electric fencing up until that point, but the bottle girls would go through it without hesitation. They would run around the front yard and come up to the house wanting to see us. And when winter came, they discovered the young, tender bark on our three-year-old fruit trees. By the end of winter, nine of the ten trees were dead because the kids had stripped off the bark.

I prefer to dam raise kids because, in my experience, they are healthier than bottle-fed kids are. The mother's milk has natural antibodies, and coupled with good barn hygiene, my dam-raised kids rarely have any problems with worms or coccidia as long as they are nursing. This is why I never wean doelings that I'm keeping. I consider it health insurance for the next generation of milkers to let them have access to all the benefits of their mother's milk for as long as the mother is willing to let them nurse, which is usually at least six months. I have occasionally heard of someone who continues giving kids a bottle for that long, but most people stop bottle-feeding around two or three months because it is a chore.

Bottle raising kids is considered a disease-prevention strategy because illnesses such as CAE, Johnes, and mycoplasma can be transmitted through raw milk and infect kids. Although testing is available, a goat may not test positive for months after infection. If an owner shows their goats, it is possible that a goat could become infected and not test positive until after infecting its kids. You can also bring one of these diseases into your herd by purchasing an infected animal. Under a disease-prevention program, all kids are removed from their mothers at birth and fed heat-treated colostrum and pasteurized milk.

Even if you decide to dam raise kids, there may be times when you have to bottle-feed a kid or two. Some breeds, like Nigerian Dwarves and Nubians, are more likely to have triplets or even quadruplets, but not every doe is able to produce enough milk to feed that many. Because a first freshener has no milking history, I will bottle-feed two of the kids if the doe has quadruplets. When a first freshener has triplets, I watch the kids carefully to make sure that they are all growing equally. Even if a doe makes enough milk, there is a chance that a smaller kid may not be getting its fair share. Does that freshened in the past should be able to feed triplets, but if they have more than that, previous milk records and her dam's kidding history will indicate whether it's realistic to assume that she will be able to provide enough milk for that many kids. A Nigerian has to peak at half a gallon a day minimum to feed four kids, whereas a miniature breed would need to peak at a gallon, and a standard goat would have to produce at least a gallon and a half. While kids will survive with a dam producing at these levels, they will not grow as fast as twins or singles born in the herd.

One type of kid rearing is not more time consuming than the other. You will be spending your time doing different things, though. If you bottle-feed kids, you'll be spending time making bottles, and you have to stick to a schedule for feeding the kids. If you dam raise, you will need to spend time with the kids daily, but you can do it whenever it is convenient for you. I usually do it at the end of the day, while I'm finishing chores, because that works for me.

Getting Started with Dam Raising

If you have decided to dam raise your kids, you'll want to make sure they know how to nurse before you leave the barn after the birth. Some kids will be up and looking for the teat within five or ten minutes of being born, but others may seem clueless. A kid that isn't looking for the teat within about half an hour of birth can be placed under the doe with the kid's nose next to the teat. That's all some kids need to wake up to the possibility of their first meal. The truly clueless kid might need to have its mouth opened and little milk squirted on the tongue to encourage it. And once in a while, you need to put the teat in the kid's mouth and support its chin as it starts to suck. After nursing once, most kids know what to do. Don't be alarmed when you see the kids nursing for only thirty seconds or a minute. That's perfectly normal because goat babies nurse very frequently.

Getting Started with Bottle-Feeding

Starting to bottle-feed is similar to starting the kid nursing in dam raising in that some kids will catch on quickly while others will fight as if they think you're trying to poison them. In most cases, you will need to open the kid's mouth and put the nipple in it while supporting the chin. At some point within a few feedings or a few days, the kid will start to grab the nipple as soon as it sees it.

Kids should have their first colostrum as soon as possible after birth. The textbooks say they should consume 10 percent of their body weight within the first twenty-four hours. Many kids will happily consume more, especially when they are receiving five feedings. To determine how

Bottle-feeding Standard Goats

Age	Amount of milk	Times per day	Total
1–3 days	3–4 oz.	5	15–20 oz.
3–14 days	8–12 oz.	4	32–48 oz.
2 weeks–3 months	16 oz.	3	48 oz.
3–4 months	16 oz.	2	32 oz.
4 months–weaning	16 oz.	1	16 oz.

Bottle-feeding Nigerian Dwarf Goats

Age	Amount of milk	Times per day	Total
1–3 days	1–2 oz.	5	5–10 oz.
3–14 days	2–4 oz.	4	8–16 oz.
2 weeks–3 months	8 oz.	3	24 oz.
3–4 months	8 oz.	2	16 oz.
4 months–weaning	8 oz.	1	8 oz.

much colostrum to feed, calculate the weight of the kid in ounces. For example, a 7-pound kid is 112 ounces, and 10 percent of that is 11.2 ounces of colostrum in the first twenty-four hours. A 3-pound Nigerian kid is 48 ounces, so it would need 4.8 ounces of colostrum. The charts on the previous page are guidelines for how much milk a kid needs and how often. If a kid develops diarrhea, reduce the amount of milk per feeding but increase the number of feedings per day so that the kid still consumes the same total per day. If the diarrhea doesn't stop within a few hours of reducing the milk per feeding, check for other causes of diarrhea.

Kids can be fed using a variety of goat and sheep bottles and teats, and I've even used human baby bottles, which are the least effective as the hole is too small and the nipple too short for some kids. Lamb and kid teats are longer and do a better job of stimulating the sucking reflex in kids. If you are going to purchase a livestock nipple that does not have a hole in it already, you should buy several because it is very easy to cut

Even if you plan to dam raise, you may find yourself bottle-feeding if a doe dies or has more kids than she can feed, so it's a good idea to have supplies on hand for emergencies.

a hole too large for a newborn. Cutting the opening on a nipple is not an exact science. It is best to make a very small opening initially. It can always be made bigger. If you accidentally make the opening too large, don't throw the nipple away. You will probably be able to use it when the kid gets older. At some point, the nipples will start to deteriorate. When a nipple feels sticky, it is time to throw it away because it could fall apart and create a choking hazard for the kid.

Goat milk, cow milk, or milk replacer

There is probably no question that will elicit more passionate responses from goat breeders than asking them what to feed kids. Most breeders have had a bad experience with bottle-feeding at some point, and they blame whatever they were feeding. One website will advise feeding milk replacer, but another will claim that milk replacer kills kids, so you should feed cow milk. The bottom line is that when you are not feeding raw goat milk, there could be digestive disturbances that are caused by what you are feeding. In our early days of raising goats, when we didn't have goat milk to feed bottle babies, we sometimes used store-bought cow milk, and sometimes we used milk replacer. The only time we had a tragic situation was when we fed a multi-species milk replacer. We happened to have a couple of bottle-fed piglets at the time, and it seemed easier to buy just one milk that would work for both the piglets and the doeling. The day after we opened a new bag of the milk replacer, both piglets and the doeling died, leading us to believe that the bag was contaminated.

If you are using colostrum or milk from does that have tested positive for diseases such as CAE, the milk should be pasteurized by heating it to 145°F for 30 minutes. Colostrum is heat treated rather than pasteurized because it will turn into curdled pudding if heated up too much. Treat colostrum by heating it to 135°F for 60 minutes.

Poop

Newborn poop sometimes causes concern for goat owners. As with human babies, the first stool will be black and sticky. By the end of the first day, it will be the color of mustard or the yolk of a pastured hen's egg. It may be the consistency of scrambled eggs or a little looser. If it looks like someone has squirted mustard on the back end of a kid in the first day or two, it means the kid is getting too much milk. We've had this happen a couple of times, and completely milking out the doe corrected the situation. A doe's udder may be overly full by the time she kids, and if the doe is especially patient, she may allow a kid to nurse too long and overindulge. A streak or two of blood appearing in otherwise normal poop in the first few days is not a cause for concern.

Feeding Grain and Forage

When a kid is dam raised it will imitate its mother and be nibbling at hay and grain within a few days of birth. When a kid is bottle-fed, it tends to start solid food much later, but you should provide hay by two weeks of age. Housing bottle-fed kids with other kids the same age may encourage them to start eating solid food sooner.

Kids still need milk or milk replacer after they start eating solid foods. Early weaning can result in kids that grow slowly and are more sickly than those weaned later. Milk and milk replacers have far more protein and calcium than forage or grain. It simply isn't possible for kids to get enough nutrients without milk until they are at least two months old.

Horns

I had failed to do my homework before bringing home my first goats. One of the unhappy surprises I had was learning that dairy goats are born with the ability to grow horns and that most dairy goat breeders disbud kids to stop the horn growth. I was not happy about the prospect of burning my kids' skulls for what I initially thought was merely a cosmetic issue. Then someone gave me two horned goats, and I quickly learned why so few people want horned goats. After three months I called the person who had given me the goats and told her I couldn't keep them.

Goats with horns can be a danger to humans, other goats, and even to themselves. One day when my husband picked up one of the horned goats, it jerked its head backwards, barely missing my husband's eye as the horn poked his cheek. When the horned goats butted heads with disbudded or polled goats, the horned goats had the advantage, and the goats without horns appeared to be in pain as they stumbled backwards. The final straw came when I saw the wether hooking his horns under the belly of a pregnant doe and lifting her off the ground. That was the end of horned goats on our farm. Horned goats are also notorious for getting their heads stuck in fencing, putting them at risk of a broken neck and making them easy targets for predators.

Disbudding

Disbudding goats is best done with a disbudding iron, which burns the horn buds and stops them from growing. For best results, bucks should be disbudded within the first week after birth, and does by two weeks. The longer you wait, the larger the horn bud grows and the longer you will have to burn to remove it. This is not fun, but procrastinating will make it worse.

You really should see an experienced person disbud goats before trying it yourself. Some breeders are willing to let a new goat owner watch them disbud kids. If you don't have a breeder near you, large animal vets can also disbud kids, although be sure they have experience with goats. Disbudding cattle is far more forgiving because cows have much thicker skulls.

There is a multitude of online videos showing disbudding, but not all of them offer good information, in particular with respect to the depth of the burn. It is possible to burn through the skull, which will result in death, of course. The successful burn is not deep, but it is wide enough to cover the full horn base. Scurs, which are tiny bits of horn growth, may grow in places that were not burned. Doelings tend to have small horn bases, making it easy to burn everything. Bucklings, however, have wider horn buds, which means it is easier to miss a part.

There are disbudding irons with interchangeable tips and with a non-removable tip. I prefer to use an iron with a non-removable tip because this type of iron will get hotter than the type with interchangeable tips. Regardless of the size of goats you have, the tip for the standard goat works best. The tip for dwarf or pygmy goats is not large enough to do a good job on smaller goats. Irons vary with respect to pre-heating times and how hot they get. Some irons get much hotter than others, and some take so long to get hot that users don't wait long enough for the iron to get hot enough to do a good job. Whenever strict times are suggested for pre-heating and for burning, make sure you know the make of the iron being used.

Although there are other ways to disbud horned cattle, for example using caustic paste or a "scoop," they are not recommended for goats. Caustic paste takes at least half an hour to work, so instead of a kid screaming for a few seconds, it will scream for much longer as the acid slowly burns the horn bud. There are also stories of kids rubbing their head against something and smearing the paste into their eyes, causing blindness. Scooping out the horn bud on goats is not recommended because the skull of a goat is much thinner than that of a cow, so the risk of death is greater.

Scurs

If you miss part of the horn bud when disbudding, the kid will grow a scur. It is part of the horn, and unless a large portion of the horn bud was missed, scurs are usually small. Because bucks have a wider horn base, they are more likely to have scurs, and because intact bucks have testosterone, their scurs tend to grow longer than those of wethers. Unless you use a disbudding iron that is too small or malfunctioning, it is unlikely that you will have scurs on does.

Scurs are usually nothing more serious than a cosmetic issue. The tip of a pointy scur can be rounded off using hoof trimmers so that the scur doesn't hurt you if the goat rubs its head against your leg. Scurs usually have a small blood supply and might bleed if you cut too close to the base. When goats butt heads, they can knock off a scur, which will also cause bleeding, which usually looks worse than it is. You can clean it with hydrogen peroxide, but the goat won't be happy about it, so you'll need someone to help you. Be careful not to get the hydrogen peroxide in the goat's eyes. "Gentle iodine," which is a one percent solution, is a good disinfectant for injuries like this and is likely to elicit a less negative reaction than stronger iodine.

When disbudding, you can use a special kid holding box, someone can hold the kid for you, or you can hold the kid on your leg as shown in this photo, a technique we learned from Ellen Dorsey, whose husband disbuds Alpine, Nubian, and Nigerian kids holding them like this. The kid's body is laying on the leg with its two left legs on one side of the person's leg and its two right legs on the other side of the person's leg. This is a right-handed person who put the kid on his left leg, holding it securely against his body with his left arm while holding the disbudding iron in his right hand.

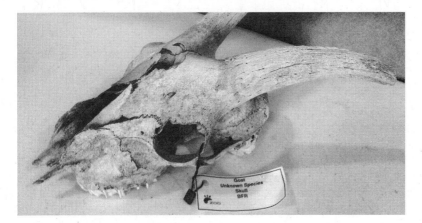

As you can see, the horns are actually part of the skull. Dehorning, which entails cutting off a horn that has already developed, is more dangerous than disbudding because the horn is hollow. The risk of infection is high, and there will be a hole in the skull where each horn was removed.

Step-by-step Disbudding

1. Find the horn buds. Although you can't usually see the horn buds through the kid's hair, you can feel them. They are simply a raised part of the skull. If you shave the hair from the kid's head, you will be able to see the horn bud somewhat. Shaving the hair off using dog clippers with a #10 blade before disbudding is a good idea because it makes the horn bud easier to see, and burning hair is smoky and smells terrible.

2. The disbudding iron should be heated up until it is literally red hot. The hotter it is, the more quickly you will be able to do the job. An iron that is not hot enough will require more contact time on the kid's skull, increasing the risk of overheating the brain.

3. The disbudding iron is placed over the horn bud for a few seconds at a time until you see a copper ring where the iron was in contact with the head. Some sources on the Internet suggest counting to ten, but we never burn for more than three or four seconds at a time. You can always burn a second, third, or fourth time if you don't see a copper ring, but if you burn through a kid's skull, you don't get a second chance. It is also possible for a kid to develop a form of encephalopathy if its brain overheats during the disbudding process.

4. Once you see a copper ring, the skin in the center will start to separate. Use the edge of the disbudding iron to flick the "cap" off the horn bud.

5. After removing the cap, turn the disbudding iron to the side and burn the middle of the horn base until it is also copper colored. You may or may not see a small amount of blood. If you do see blood, use the iron to burn the spot to cauterize it and stop the bleeding.

6. If kids are dam raised, you should immediately stick the disbudded kid under its mom to nurse. You will notice that she will sniff under the tail to make sure it's hers. Although does and kids recognize each other's voices, smell is the litmus test, and if a kid doesn't smell right, some does will reject it. After disbudding the sooner you put the kids back with mom, the lower the risk for rejection. Although it is usually temporary, it is worrisome when a doe won't let her kid nurse.

Polled goats

Polled goats never grow horns. No disbudding is required. When a polled goat is bred to a genetically horned goat, half of the offspring will be polled. The polled gene is dominant, which means that if a goat has a gene for polled, it will be polled. If a goat was born with horns, it has two genes for horns. If it was born polled, it most likely has one horned gene and one polled gene. Most breeders in North America don't breed polled goats to each other because of the increased risk of hermaphroditism, and for this reason. you won't find breeds of polled goats, even though there are breeds of polled cattle.

Polled Goats— Types

Heterozygous polled. A goat has one polled gene and one horned gene. If either of its parents was genetically horned (born with horn buds), a polled goat is heterozygous.

Homozygous polled. A goat has two polled genes. Both parents must have been polled. There is no test for this in goats. A goat that has thrown at least ten kids, all polled, is likely homozygous polled. Even though a homozygous goat can throw only polled kids, if it is bred to a horned goat, the kids will be heterozygous polled because they will have a horned gene from the horned parent.

The polled gene in goats is not fully understood, but in research conducted in the early to mid twentieth century, several studies showed a definite connection between breeding polled-to-polled parents and hermaphrodite kids. In one study, when two heterozygous polled goats were bred to each other, 10 percent of the kids were hermaphrodites. When a heterozygous polled goat was bred to a homozygous polled goat, 25 percent of the kids were hermaphrodites.[14] Polled goats almost became extinct as a result of such studies. However, with the increased interest in raising dairy goats recently, there is a renewed interest in polled goats because new goat breeders tend to greatly dislike disbudding. There is also a very small number of people starting to breed polled goats to each other in the hope of getting homozygous polled bucks.

The arrow indicates the swirl of hair that grows around the horn bud on the kid's head.

If you have a kid from a polled parent, you should be absolutely sure it has horn buds before disbudding. It has been argued that the polled gene can skip a generation, evidenced by disbudded goats that gave birth to polled kids in spite of the fact that the polled gene is considered a dominant gene. However, there is always the possibility of human error—the breeder thought they felt horn buds and disbudded a goat that was genetically polled. Whether the gene can skip a generation will remain a matter of speculation until we have a test for the polled gene in goats. Because I never want to get an email from a buyer telling me that the disbudded goat I sold them is throwing polled kids, I tend to wait a little longer with kids from polled parents before disbudding to be absolutely sure that they are indeed growing horn buds.

To determine whether a kid from a polled parent is horned, you need to check its head daily from the time it's born. Most bucks have little points on the top of the head the day they're born. In every case except one on our farm, when a buckling did not have horn buds at birth, it was polled. I got lucky with the kid that was horned because I had a buyer who had horned goats and only wanted the buck if he was horned, so I know we didn't make a mistake and disbud a polled goat. Does can be a little more challenging because most of them do not have noticeable horn buds until they are at least a few days to a week old.

Polled kids will get bumps on the top of the head, which makes determining whether there are horn buds a challenge. You may be able to feel bumps on the top of a polled buck's head a few days after birth. How do you tell the difference between horn buds and polled bumps? First of all, polled bumps are smooth, or rounded, across the top, whereas horn buds are pointy. The points of horn buds will become more evident each day. The polled bumps will grow out, getting wider, whereas the horn buds get pointier and pointier until they're like the point of a pencil. Horned kids have a swirl of hair around each horn bud, whereas the hair on the head of a polled kid grows from the center of the head towards the edge with no swirls around the poll bumps. Although it is subjective, a polled kid's head is said to be more egg-shaped than the head of a horned kid. With time and experience, you may find yourself saying a kid is polled before even touching its head.

Castrating Males

It is a simple fact that you don't need very many bucks for a dairy herd. Because a buck can sire dozens of kids, you should keep only the best for breeding. That means that a lot of bucklings will become pets, brush eaters, or meat. Unless they will be butchered in a few months, bucklings should be castrated because intact bucks get stinky and pee on themselves. They also tend to fight with each other during the breeding season. There are three methods of castration, and breeders can easily learn the methods themselves.

Banding

Banding is probably the most popular method of castration because it is simple and inexpensive. A rubber band the size of a penny is placed around the base of the scrotum using a special tool that opens up the band wide enough to get it over the testicles and in place. Some argue this is the most inhumane method of castration because it cuts off blood flow to the entire scrotal area, which causes everything below the band to atrophy and fall off. We used this method for a few years, and most

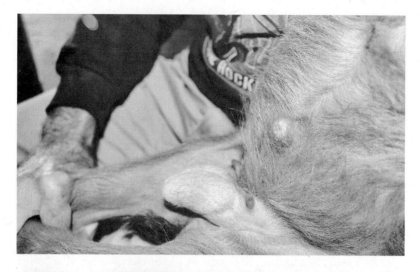

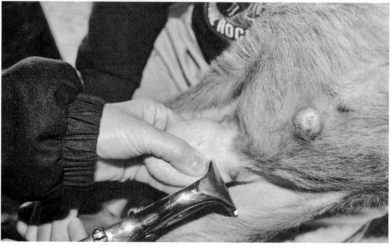

Castration usually requires two people, one to hold the kid and one to do the castration. The cord that leads to each testicle is clamped separately. I hold the cord in place to make sure it does not slip out of the burdizzo as I am clamping it.

bucklings didn't seem terribly bothered by it. A few bucklings would scream for a few minutes up to an hour, and some would get very depressed for a few hours or a day following banding. Because of the anaerobic environment that exists under the band, there is a risk of tetanus with this type of castration.

Emasculator

Emasculation is the safest method of castration because the skin is never broken, but it is not immediately obvious that you have done the job.

With this method the cord that goes to each testicle is crushed using a special instrument called a burdizzo. Although a kid usually lets out a short bleat when the cord is clamped, most recover fully within fifteen minutes. Some cattle ranchers say they have an unacceptably high rate of failure with this type of castration, which makes sense when you see that a cattle burdizzo is quite large and requires the use of two hands to operate. The goat and sheep burdizzo is much smaller and can be closed with one hand by most people. We started using this method several years ago and have not had any failures.

Surgical castration

When I had my first goats, I read that surgical castration was the most humane method, so I took my first kids to the vet for the procedure. As we stood in the parking lot, I held the bucklings as the vet sliced open each side of the scrotum, pulled out each testicle and dropped it on the ground. He told me he was leaving the scrotum open so that it could drain because stitching it up would be more likely to result in an infection. Although the boys survived the ordeal and were just fine, I decided to look into other methods of castration. Some people do prefer surgical castration, though, and if you want to do it yourself, you should have a vet or an experienced breeder teach you.

Tattooing

Registered goats must have a form of permanent identification. A microchip can be put on the underside of the tail, but not all registries accept this. This method requires a microchip reader, which is expensive, so most dairy goat breeders tattoo their animals inside each ear. Because LaManchas don't have enough external ear cartilage, they are tattooed on the skin on the underside of the tail, usually in the loose skin on each side of the bone. Green ink works for black goats as well as for lighter colored ones.

When you become a member of one of the goat registration organizations, you will register a herd tattoo, which is usually three or four letters unique to your farm. The herd tattoo is put in the right ear of every goat that is born in your herd. The personal identification tattoo is put in each goat's left ear. It includes a letter that indicates the year the kid was born and a number that signifies the kid's birth order. For example, the fifth kid born on your farm in 2013 would have D5 tattooed in its left ear. The dairy goat registries in the United States recommend D for 2013, E for 2014, F for 2015, and so on. The letters G, I, O, Q, and U are not used because they look too much like other letters and numbers.

Tattoo pliers are available from most goat supply companies. I recommend getting the pliers with the auto-release mechanism because it removes the pins from the kid's skin as soon as you release your grip. If you don't use tattoo pliers with this feature, it can be stressful for human and goat as you peel the ear off the pins.

Weaning

On our farm, we never wean doelings that we are keeping. If they are being sold, they continue nursing until the day they leave. Some people feel that a kid should be weaned before it is sold. I don't do this, however, because I feel its health will suffer as a result. The longer a kid nurses, the better. Unfortunately, weaning and moving are both stressful events that frequently result in increased risk of parasites, coccidiosis, or even stress-induced diarrhea. When a kid moves to a new farm, it can develop these problems, even if it has been bottle-fed and does not have the stress of leaving mom. It is still leaving the only home it has ever known. Goats do not like change, and some are more bothered by it than others. For that reason, I see no point in prolonging the amount of stress that a kid will unavoidably suffer as a result of being sold. Ultimately, it is less stressful for them to simply have one bad day where they go to a new farm without mom.

Most of our goats don't get terribly upset about going to a new farm, anyway, because they have already been subjected to regular separations from mother. After two months of age, a kid spends the day with its mother in the pasture, and they are separated overnight so we can milk the doe in the morning. This nightly separation could be viewed as pre-weaning. Dam-raised kids start eating when they are a few days old, and most are eating quite well and pooping little goat berries well before they are even a month old.

Bucklings are usually separated from their dams by three months of age due to the possibility of breeding doelings in the herd. The odds of a buckling at that age breeding an adult doe are quite slim because they usually aren't big enough. You can also keep bucklings separated for most of the day and give them a couple hours with their dams daily for nursing. As long as the mother is not in heat, the risk of pregnancy is zero, even if the buckling were able to mount her.

Because the stress of weaning can subject bucklings to coccidiosis and parasites, it is best to keep them separated from adult bucks and wethers. If you are keeping only one buckling intact during a kidding season, you can give him a wethered buckling for companionship during weaning. When you purchase a buckling, it is best to purchase either two bucklings or to get a wether as well as the buckling. But since a wether eats as much as an intact buck, you might want to buy two bucks.

When weaning a doe kid or a wether, it may be necessary to keep it separated from its dam for many months before you can be sure that the kid won't nurse again. Some kids will forget about nursing after a couple weeks, but some may go back to nursing after a couple of months if the mother is willing and still in milk.

Wethers that are to be raised for meat will grow faster with the added protein and calcium in the milk if they are left with the doe to continue nursing until you are ready to butcher them. A buckling being raised for meat will grow faster if left intact. Does of the Swiss breeds don't usually cycle during the spring or early summer months, so you can probably keep a Swiss breed buckling intact and nursing for a few months, getting the highest meat yield.

Barn Hygiene

When I was new to goats and heard that you had to keep the barn clean to avoid coccidiosis in kids, I thought that was unrealistic. How can you keep a barn clean? It isn't as difficult as it sounds. My benchmark is simple: if I'm not willing to sit down somewhere to play with kids, then it isn't clean enough for kids. Just as human babies put everything in their mouths, so do baby goats. The more exposed manure there is, the greater the chance that a kid will check it out at some point by picking up manure with its mouth or nibbling at a piece of straw with manure on it. During the summer stalls need to be cleaned out regularly by removing all of the bedding and manure and replacing it with clean bedding. During the winter a thin layer of bedding is added regularly so that there isn't any exposed manure. You may need to do this somewhere between daily and weekly, depending on how many goats you have in your space.

CHAPTER 12

Milking

✦ ✦ ✦

I F YOU ARE bottle raising kids, you start milking the does as soon as the kids are born, and you continue milking twice a day, every day, until you dry them off, although you should be checking her udder daily from day one to be sure both sides are even. If you are dam raising the kids, the management is a bit more flexible, and it will free you from daily milking at least sometimes.

Managing Milkers Naturally

If a doe has a single kid, it is ideal to start milking her three or four days after she gives birth. A doe produces milk in response to demand, and if only one kid is nursing, the result often is a fat kid and a low milk supply. By regularly milking a doe with a single kid, you will also avoid the possibility of her getting a lopsided udder if the kid decides it has a favorite side. You can put her on the milk stand twice a day, every day, without separating her from the kid. Or if you prefer to milk only once a day, you can separate the kid from her overnight and milk her in the morning. Separating the kid and doe by putting the kid in a dog crate in the stall with the doe reduces the stress of separation for both animals because they can still see each other. I usually start out separating them for only eight hours and gradually build up to twelve hours over the course of a week.

A doe with twins can be put on the milk stand once or twice a day and milked. If the kids are with her all the time, they can consume as much milk as they want, so you are only taking the excess. And remember, demand creates supply. You can start separating the doe and the kids overnight a couple of times a week when the kids are two weeks old. Once you know how much milk the doe is producing, you may be able

What I learned from Showgirl

When we first started showing our goats, most of the other breeders I met at shows were bottle raising their kids, and over the years many asked me the same question—don't you have trouble with lopsided udders when your does raise their own kids? No, we never had a problem. Since this seemed to be such a big deal with other breeders, I came to the conclusion that my goats were just smarter than most! Then, after seven years, I was humbled.

I stood in the barn in complete shock as a yearling doe walked away from me and appeared to have a one-teat udder. I knew she had two teats, but it looked like she had one teat hanging down from the center of her udder when I looked at her from behind. I quickly ran towards her and felt her udder, immediately finding a second teat flat against her abdominal wall. I knew what had happened. She had freshened a week earlier, and apparently Showgirl, her doeling, had been nursing on only one side, causing the other side to dry up.

The lesson was reinforced a couple weeks later when another doe gave birth to a kid that favored one side. But why hadn't this been a problem previously? Well, we started out always milking does with single kids because we wanted the milk. As our herd grew and our supply of milk increased, we were no longer faithful about milking the single kid does. A kid can wind up with a favorite side shortly after birth if it nurses on only one side. That side gets soft, which makes it easier to nurse, so the kid starts avoiding the fuller, harder side, and within a few days, the other side starts to dry up.

to separate her more often. I had a LaMancha that peaked at 2 gallons a day, so we could have separated her every night and taken a gallon, and her twins still would have had a gallon of milk during the day. A doe that is waking up her kids often and pushing them towards her udder is probably uncomfortably full, and it is a good idea to milk her at least once a day.

If a doe has triplets or more, I let her nurse them exclusively, and I don't start milking for the first two months other than our once-a-month milk test. If I have an especially heavy producer feeding chubby triplets, I may start separating them overnight a few times at around six weeks.

You can separate kids overnight and milk the dam in the morning, leaving them with her during the day, for as long as you want. If we want to make a lot of cheese, we separate the kids for a couple of days. On the other hand, if we will be going somewhere early in the morning, we may not separate the kids overnight so that we have fewer goats to milk before we have to leave. Most kids are remarkably good at keeping the mother's udder close to empty. When we first start separating the kids overnight, we put each doe on the milk stand in the evening to be sure she doesn't have a lot of milk, and in most cases, there are only a couple of squirts. If we discover a kid isn't taking all of the milk during the day, we milk that doe in the evening to keep her supply up.

Every few years I have a doe that will refuse to let a kid nurse after a sibling leaves, or a kid that stops nursing after a sibling leaves. For this reason, you should put a doe on the milk stand twice a day for the first few days after making any changes such as selling a kid or moving a buckling to the weaning pen. You might also discover in the case of twins that each kid had a habit of nursing from only one side. If that's the case, you will need to milk the absent kid's favorite side daily to avoid the possibility of one side drying up.

Teaching a Doe to Milk

Teaching a doe to be a well-behaved milker begins long before she even gets pregnant. When we had only a few goats, none of the does ever received any grain unless they were on the milk stand. Now that we have a larger herd, we are not quite that organized, but the basic idea is that you should be interacting with the does on a regular basis from the time they're born so that they trust you. If you don't put them on the milk stand on a somewhat regular basis, avoid doing it only for unpleasant experiences such as deworming, injections, and hoof trimming. When a doe associates the milk stand with only those activities, it will be much harder to gain her trust when it is time to milk her.

There are a number of different ways to acclimate young does to milking. In some cases, a doeling will follow its mother into the milking parlor and nibble on grain or hay while its dam is being milked. The doeling learns from her that the milking parlor is a great place to be! Unless I need to milk her sooner, the morning after a young doe freshens, I lead her to the milk stand to give her grain. If she doesn't already know the routine and jump on the milk stand on her own, I hold the pan of grain under her nose and let her get a bite or two. Then I start moving it towards the milk stand, finally holding it above the milk stand to en-

---

✢ MARIN WADDELL, SalayView Farm, Lang, Saskatchewan

We have a fairly new small farmstead creamery. Our first year in production we milked twenty Nigerian Dwarf goats and three LaManchas. Our main goal with our herd is milk production for the business, so we decided to take most of the kids away from their mamas when they were young. We bottle-fed them so we could use the milk for cheese. Our pasteurizer has a 100L (26 gallon) minimum so we wanted as much milk from them as possible.

We started the kids off on individual bottles with Pritchard nipples and then transitioned them to lambars after a week or so. In addition to colostrum in the beginning, all bucklings were given milk replacer, and doelings were given mostly raw goat milk with a top-up of replacer.

We did it this way because we knew that raw goat milk is best for growing healthy kids. We raise the doelings to add to our milking herd; bucklings go for meat. We also had several does kid later in the season and, rather than starting up the bottle-feeding again, we chose to leave those kids on their mamas and separate them at night, milking the mama once a day in the morning.

Separating kids completely before weaning age is not something that we will do again. There are several reasons. The first was that the kids weren't as healthy as the ones who stayed with their mamas. They didn't grow as fast, and we had issues with scours, most likely coccidia because it cleared up when treated. The

courage her to jump up there. Some does follow along eagerly, but with others it might take a few minutes. Even if I don't want to milk the doe, I gently handle the udder and teats while she is eating the grain. If the udder has blood on it from the birth, this is often a good time to wipe it off using a warm cloth. I continue giving the doe grain on the milk stand once a day for the first week or two after she's kidded, or at least until she runs for the stand and jumps up on it like an old pro.

The sooner you start milking a doe after she kids, the more likely it is that she will be cooperative, especially if she is a first or second freshener. Since singles are fairly common for first fresheners, and does with single kids should be milked daily, it often works out well. Goats thrive on routine, and if you routinely milk them when they are nursing kids, they will view it as normal. Most experienced milk goats that have freshened a few times are willing to let you milk them when they are nursing kids.

second reason was the amount of work. Preparing formula and then cleaning up the buckets and nipples afterwards was annoying. The third reason was milk volume. Although we did get more milk from the does because we were milking them out twice a day, their overall production was less than the does whose kids were left on them, plus we were feeding some of the does' milk back to their doelings. Getting the extra bit of milk was not worth the impact on goat health or the increased workload for us.

This coming milking season we will start separating the kids at night as soon as we feel is reasonable, depending on size of the kids and number per litter, and we will milk just once in the morning. This might give us a slower start to our cheese-making season as it will take a bit longer to collect the 100L volume requirement, but the payoff will be in more milk throughout the season and healthier kids.

We're still relatively new to all of this, and we'll continue adjusting our feeding/milking program as necessary because having healthy animals is an integral part of having a healthy business. Plus we love our goats and want to do right by them.

If you wait a few weeks to start milking, it could be more challenging. Kicking and lying down are the two most common problems people face when milking new goats. Using hobbles on a standard goat is a common solution for kicking. If you have Nigerians or miniature dairy goats, you might be able to hold one of the hind legs in the air or have a helper hold the goat's hind legs while you milk. Holding one leg up usually works better than trying to out-muscle the doe and hold both legs down. If a doe lies down on the milk stand, you can put something tall under her chest, or again, if it's a Nigerian, someone might be able to hold up her back end while you milk. Most does will calm down for milking somewhere between a day and a week after you start milking them.

Milking Equipment

You will need the following items whether you will be milking by hand or with a machine. All of these things can be purchased through goat supply companies, but in some cases, you can make your own or come up with creative substitutes that cost less.

Milk stand

Don't skip this one! Someone once bought a perfectly trained milk goat from me and then complained that she was having problems milking the doe. After a discussion, I realized the woman did not have a milk stand. There are a few sweet goats in this world that will let you milk them anywhere, but you need to use a milk stand for most goats, especially if you are a beginner. Trying to milk without a stanchion will lead to fighting with the doe, which sets you up for failure when you put her on a milk stand because she's had a negative experience with milking. Goats have great memories, so you want to start out right.

You can buy a fancy metal milk stand, or you can make a wooden one using scrap wood you find around your farm. Commercially made metal milk stands are nice because the deck is a metal mesh, so when you spill milk on it (and you will spill milk), the milk stand doesn't become slippery. Wood milk stands get very slippery when they're wet. The big difference between the two, of course, is price. A professionally made

We made this milk stand from scrap wood, and we are still using it eleven years later. There are a multitude of designs for milk stands, but the important thing is that there is a place for a feed pan and a way to hold the goat's head in place so that she can't decide to leave when she's done eating. After a couple of years milking in an open barn, we built a milking parlor to keep our equipment cleaner and more organized.

metal one is a few hundred dollars, whereas you can make one for a fraction of that price or even free if you have some spare wood.

The height of the milk stand is important. It should be short enough for the goat to jump up on it but high enough for you to sit comfortably next to it and milk the goat. Commercial milking parlors usually have very high milking stands, which goats can access by walking up a ramp, and the humans stand behind the does to connect the inflations from the milking machine.

Milk bucket

The bucket should be stainless steel and seamless for ease of cleaning. You don't want any cracks or seams in the bucket where bacteria can hide. A 6-quart bucket works well for most standard goats, but you'll want a shorter one for Nigerians. Milking machines do not get the udder completely milked out, so a milk bucket is still needed for hand milking to empty the udder.

Strip cup

An official strip cup is a stainless steel container with a wire mesh filter that sits on the top. The filter alerts you to chunky milk (a symptom of mastitis) before you start milking. However you can use a repurposed tin can or an old coffee cup. The strip cup is used to collect the first few squirts of milk from each teat. Research has shown that the first squirts contain a larger amount of bacteria. We give this milk to the barn cats after milking.

Udder supplies

You can buy disposable udder wipes or baby wipes for cleaning the udder. Some people use a bucket of warm, soapy water and actually wash the udder. We simply use a warm washcloth. Whatever cloth or wipe you decide to use, a clean one must be used for each doe so that you don't spread germs.

Teat dip

Iodine is used in organic dairies to dip teats, but there are also chemical teat dips and sprays available. If you use a dip, you'll need something to hold the dip. Old-fashioned film canisters or prescription pill bottles work well. If you will be letting the does out to be with their kids after milking, you don't have to use a teat dip because the kids will be nursing through the day. The purpose of the teat dip is to sanitize the end of the teat as it closes up during the fifteen or twenty minutes following milking. This doesn't happen when a doe is nursing kids because kids nurse so often. The frequent nursing reduces the risk for infection as it keeps milk flowing out of the teat.

Milk filters

Goat hair and dust will inevitably wind up in your milk bucket, so you'll want to filter the milk before storing it. Filters are disposable. Cheesecloth stretched over the top of the bucket while milking also works as a filter, but you'll need to wash and boil the cheesecloth between uses to keep it sanitary.

Storage containers

You probably already have plenty of things in your house to use for milk storage, such as a pitcher with a lid or canning jars with plastic lids. We

Milking a goat by hand involves trapping the milk in the teat and then squeezing the teat so that the milk squirts out.

also use carafes and old-fashioned milk bottles and cover the tops with aluminum foil. If you don't cover the container, the milk will develop an off taste and a dry film on the top.

Milking by Hand

If you have only a few goats, you might be able to milk by hand. My arms are burning by the time I finish milking one standard-sized goat or about five Nigerians, but my youngest daughter has milked as many as sixteen Nigerians and two LaManchas by hand, so ability varies tremendously among individuals. If you have health issues like carpal tunnel syndrome or arthritis in your hands, you should probably look into getting a milking machine.

When a doe has long teats, the milking method uses four fingers: the index finger wraps around the top of the teat and squeezes, then the middle finger wraps and squeezes, then the ring finger, and finally the little finger. If nothing comes out after doing this several times, you may not be squeezing hard enough. If you feel the teat get smaller as you squeeze, but nothing comes out, you are not trapping the milk in the teat, and it is going back up into the udder. When a doe doesn't have long teats, you may not be able to use all four fingers, and if she has really short teats, you may simply need to roll your thumb against the teat (from the udder down) while pressing it against your index finger. Gently massaging the udder and placing a warm cloth on the udder will help the goat let down her milk. Some people bump the bottom of the udder like kids, although definitely not as hard as kids do.

Keeping a doe happy on the milk stand usually means giving her grain or other feed. When you are learning to milk, it is easy to overfeed grain, resulting in the doe developing diarrhea in the short term and becoming obese in the long term. To avoid these problems, you can mix alfalfa pellets into the grain ration. I even do this with some of my does that are especially fast eaters.

When a doe does not have kids nursing, you need to be sure to empty the udder. Towards the end of milking, as the squirts of milk get smaller, a gentle massage of the udder usually produces a little more milk. If you don't empty the udder at each milking, production will gradually go down. This is why I recommend that people who are new to milking goats milk a doe that has at least one kid still nursing. If milking by hand doesn't empty the udder, the nursing kid will.

Milking With a Machine

There are a variety of electric milking machines and manual milkers on the market, varying in cost from a hundred dollars to a thousand or more. The less expensive machines are manually operated and milk only one teat at a time, so they are not really practical if you have more than a couple of goats. Electric milking machines milk both teats at the same time and can be configured to milk multiple goats if you have a large herd.

There are a few things that are always important when using a machine. Sanitation cannot be ignored. Unless you are a really slow hand milker a milking machine doesn't save time because of the cleanup involved. All the pieces of the machine that came in contact with milk need to be disassembled and thoroughly washed and sanitized after every milking. The tubes and inflations must be cleaned with a special

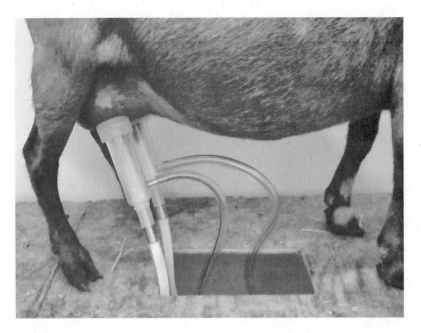

A milking machine with clear inflations is nice so that you can be sure the teat is correctly placed in the inflation.

brush, which is usually available from the companies that sell milking machines.

You will probably have at least one doe that produces milk faster than it can flow through the tubes, which means that the milk will be pooling around her teats. Unless she is the first goat to be milked, her teat orifices are now being exposed to germs from every goat that was milked before her. If you're milking with a machine, you want the doe's teats to be as clean as possible when you put the inflations on them to avoid cross-contamination between does. A pre-milking teat dip will help prevent infection.

The doe must let down her milk before you put the inflations on the teats, so put a few squirts into a strip cup after cleaning the udder and dipping the teats. Some goats are not terribly fond of milking machines and don't let down their milk very well, even if you have manually re-moved a few squirts. Massaging the udder a little after connecting the inflations will help get the milk flowing.

If the inflations are properly placed on the teats, they will stay on without your holding them. The first time you use a milking machine on a doe, she may be apprehensive, but if she is extremely upset, you may be doing something wrong. Your machine should come with information about the correct pressure of the vacuum. If it is non-electric, you need to create a suck-release rhythm that mimics a kid nursing. Never pump up the pressure and hold it! When the stream of milk starts to slow down, you can massage the udder gently to get a little more milk before turning off the machine and stripping the last bit of milk by hand to empty the udder.

Handling Milk

One of the first decisions you have to make when you have your own dairy animals is whether to consume the milk raw or pasteurized. Although some people are purists on one side or the other, using only raw or pasteurized for everything, we've decided to use both. We choose whichever form yields the best quality in the final product. When we find no difference in quality, we use raw. There are convincing arguments on the raw versus pasteurized debate, and as I've said with other decisions you'll have to make, you should do whatever helps you to sleep at night.

The cleaner the milk is, the better it will taste. When we started milking our first goat, sometimes the milk tasted great, and sometimes it tasted goaty. I tried everything that I'd read, such as rapidly cooling the milk by putting it in a jar in an ice water bath, but nothing seemed to work consistently. It was very frustrating because it meant that the quality of our yogurt and cheeses was often unpalatable. When I first started milking, however, I would just stick a bucket under the goat and start milking. After a few months, I heard that one should clean the udder, so I started using a washcloth to wipe off the teats and udder. After a few years, I read that there were germs in the opening of the teat and that the first few squirts of milk should be put into another container, so we started using a strip cup. I eventually learned that nothing can make up for starting with unclean milk.

The other important point is that when you are consuming the milk, you should have healthy animals, regardless of whether you plan to pasteurize. Although pasteurization should kill most pathogens, there is some controversy over whether it kills Johnes. CAE is not thought to infect humans, but there is the caveat that more research is needed in this area. And as already discussed, we do know that listeria, TB, and brucellosis infect humans through raw milk.

Storing Milk

Milk should be filtered, chilled as soon as possible, and kept refrigerated until ready to use. When hand milking, I prefer to filter the milk as I milk each goat, but one of my daughters brings all of the milk into the house at once and filters it then, and we have not seen a difference in the quality of the milk. When I use the milking machine, it is simpler to strain the milk after bringing it into the kitchen.

There are official storage times for pasteurized milk, but storage time for raw milk is variable. Raw milk does not "sour" under refrigeration like pasteurized milk, but storage time is more complex than that. After a few years of consuming our own fresh milk, I've started to notice a taste difference after about three days. The taste starts to remind me of store-bought milk, so we try to do something with the milk before it reaches that point. Any hint of goatiness early on intensifies with each passing day. The quality also goes down for cheese making as milk gets older. Flocculation will occur faster, and if the milk is too old, it may not be possible to get a decent curd. Once milk gets to be four or five days old in our refrigerator, I freeze it for making soap.

PART III

Milk, Meat, and More

Today our goats provide us with more than a dozen different types of cheese, and other dairy products as well as meat, fertilizer, and leather. They feed us with their milk and meat, and they also feed our pigs and poultry with the whey from the cheese that we make, and the pigs and poultry eventually feed us also. The goats fertilize the pastures where they graze, and our garden and fruit trees, which eventually provide food for us, benefit from composted goat manure. Without the goats, we would have to buy far more of our own food as well as fertilizer. After having goats for only a few years, it occurred to me that they were truly the centerpiece of our diversified homestead.

CHAPTER 13

The Dairy Kitchen

✦ ✦ ✦

When I brought home my first goats, I had no idea that I could make anything with the milk other than chèvre, but it didn't take me long to discover yogurt and many other cheeses, and I fell in love with the milk, the cheese, and the goats. When friends tried my chèvre, they started asking if they could buy it, so I called the state Department of Agriculture to ask what I needed to do to legally sell cheese. The department representative started giving me a long list of requirements, and the dollar signs started adding up in my head. I explained that I was milking only two goats, but the representative said that everyone in Illinois should have access to safe food, regardless of where they bought it.

I quickly realized that I would have to sell a lot of cheese to recover the investment of more than $100,000 to build a certified kitchen and become a licensed dairy, and then I'd have all the associated costs of running a business. And as much as I love my goat cheese, did I really want to make large amounts of cheese every day? Instead of developing a commercial dairy, I decided to learn to make every cheese and dairy product that our family uses. Because we are serious cheese lovers, I figured we could save a few dollars along the way. We have been making 100 percent of our cheese for the last few years, and we have made eighteen different types of cheese and counting.

When people hear about how much cheese we make, they find it hard to believe that we are not selling it, but I am quick to point out that aged cheese lasts forever. In 2012 a cheese shop in Wisconsin decided to close its doors and found cheddars in the cooler that were twenty-eight, thirty-four, and forty years old! Not only were the cheeses still edible, they received rave reviews. So, while it is still a good idea to contribute to a financial retirement account, we are also building up our store of aged cheeses. And because a traditional cheese cave does not need electricity, storing cheese in one is a sustainable method of food preservation.

One of the first questions I am asked about our cheese making is, "Is it hard?" Answering that question is a lot like answering the question, "Is it hard to play the piano?" It is called artisanal cheese making because it truly is an art. It is quite easy to learn the basics. Just as you can stop your musical education at any point and keep playing the same songs, you can also decide that you're happy with whatever cheeses you've mastered and stick with those forever. Or you can keep learning for the rest of your life. There are several hundred cheeses in the world to discover and master making, and you can even create your own unique cheeses.

Equipment

You may already have much of the equipment that you will need to make cheese and fermented dairy products. The specialized equipment can easily be found with online suppliers that cater to the home cheese maker.

- **Pots:** Most sets of pots and pans have at least a 5-quart dutch oven, which will be large enough to make a batch of cheese with a gallon of milk. If you don't already have a 9-quart pot or slightly larger, you may need to buy one for making the hard cheese recipes in this book, which call for 2 gallons of milk. Many traditional methods of making cheese use a water bath, which you can make by suspending a smaller pot in a larger pot that contains water that is at the same level as the milk in the suspended pot. Indirect heating is the goal, and to create it the smaller pot cannot rest on the bottom of the outside pot. This is not the same thing as a double boiler, where a pot sits above boiling water.

We are not the most orthodox cheese makers. We use thick-bottomed pots on a gas stove, and while stirring gently, we turn the flame off and on as needed to achieve the desired temperature. You should probably use a water bath on an electric stove because electric stoves are slower to respond to temperature adjustments. If you have an induction stovetop, you need the water bath because an induction stovetop gets extremely hot very quickly and the high heat can damage the milk at the bottom of the pot.

- **Cheesecloth:** This is used for draining some cheeses and for lining a cheese press when making cheeses like queso fresco, cheddar, and Gouda. A cheesecloth, sometimes called butter muslin, is available from cheese-making supply companies. An old pillowcase or other fabric that is fairly thin and has a slightly loose weave also can be used successfully. Don't buy the cheesecloth that is sold in the grocery store because the weave is too open. You will drive yourself crazy trying to get the cheese curds out from between the threads.

- **Thermometer:** Although I have made my easy mozzarella and paneer in a pinch without a thermometer, you need one to make most cheeses. A standard cooking thermometer that reads temperatures from freezing to boiling will meet your needs.

- **Spoons:** A large long-handled spoon is used for stirring the curds. It should be long enough to easily reach to the bottom of the pot you will be using.

  A large long-handled slotted spoon is used for scooping the curds out of the whey.

  You will need a set of measuring spoons, and the set needs to include a ⅛ teaspoon measuring spoon. Although you can make do with a basic set, you might want to add measuring spoons for $\frac{1}{16}$ teaspoon and $\frac{1}{32}$ teaspoon if you make mold-ripened cheeses.

- **Curd knife:** Specialty knives for cutting curds are available from cheese-making supply companies. Or you can do as I do and use a bread knife. It has a blade that is about a foot long and reaches to the bottom of my cheese pots. A 12-inch icing spatula will also work. Because curds are quite soft, a curd knife does not need to be sharp.

- **Cheese mat:** Sometimes called a drying mat, they are made of plastic mesh or bamboo. They allow the cheese to drain by creating space between the cheese and whatever it is sitting on. A cookie cooling rack won't work for cheese because the spaces are too large and the cheese will slip through. I usually put a plastic drying mat on top of a cooling rack that sits on top of a baking pan that catches the whey.

- **Molds:** Each type of cheese has a traditional shape, but in some cases, such as chèvre, you can be creative and use whatever shape you want. In other cases, such as when making Camembert, you will want to use a mold that will result in a cheese that is no more than an inch or two thick if you want the traditional gooey inside.

- **Cheese wax:** Some aged cheeses, such as cheddar or Gouda, are covered with wax to protect them from mold during the aging process. It might be tempting to put your cheese into a plastic bag and use a vacuum sealer, but this method is not a good option for long-term aging. We have found that cheddar does well for six to eight months in plastic, but after that, it starts to get soggy. Cheese wax breathes and allows the cheese to lose excess moisture during the aging process. Cheese that is sealed in a plastic bag winds up with a surprisingly large amount of liquid on its surface.

- **Cheese press:** Making most aged cheeses requires the use of a cheese press to force excess moisture out of the cheese. Choose a press that has a pressure gauge because the recipes call for the application of specific weights. Using too much weight will create a drier cheese, and using too little weight will result in a wetter cheese.

- **Cheese cave:** Okay, so a cave isn't exactly a piece of equipment, but it can be. For centuries people aged cheese in real caves or cellars where the year-round temperature was in the low to mid 50s. A root cellar where temperatures stay in the lower 50s year-round might work, or perhaps you have a corner of your basement to use for aging your cheese. Since most people today don't have a proper cheese cave, they use a separate refrigerator. A wine refrigerator can be set to the correct temperature, but a typical refrigerator cannot be set above 40°F, so it needs the addition of a special electrical device that plugs into the refrigerator and will allow a higher temperature than is safe for regular food storage. Whatever arrangement you decide on for a cave, do not start making hard cheeses until you have a proper place to age them. Even if you do everything else right, your cheese will be a flop if it is not aged properly.

- **Cream separator:** We lived without one of these for many years. You need it only if you want to make butter or if you want skim milk. Because goat milk cream does not rise to the top very quickly, as it does with cow milk, you will need a cream separator if you want fresh cream. It takes days for the cream to rise to the top, and by then, the flavor of the milk has started to deteriorate.

Ingredients

It is almost magical that the tiniest addition of acid or of culture and rennet will create a multitude of different cheeses from milk. However, without the right ingredients, your cheese is likely to fail. And while I almost never admit defeat in the kitchen and have been known to fix dishes that seemed beyond hope, failure in making cheese usually means it's a good day to be a pig on our farm. For this reason, I recommend using commercial cheese-making ingredients, which are available by mail order from reputable companies.

- **Milk:** This ingredient may seem obvious, but the real question is whether to use raw or pasteurized milk. A lot of information is available on the risks and the benefits of consuming both raw and pasteurized milk, and I'm not advocating either one unequivocally. As I've said on other topics, you need to make the decision that will help you sleep at night. Although we consume our milk raw, we pasteurize milk to make some of our dairy products because the pasteurized milk gives us consistent results. The recipes in this book do not specify raw or pasteurized milk, but keep in mind that the FDA recommends pasteurized milk for all dairy products except cheeses that are aged at least sixty days.

  When pasteurizing milk to make cheese, the milk must be heated to 145°F for 30 minutes. If you are pasteurizing milk to make any of the culture-ripened cheeses, the milk should not be heated above 170°F or you may have difficulty getting a firm curd. If you accidentally overheat the milk, you can go to plan B at that point and make an acid-ripened cheese. When cooking with milk, if a recipe calls for milk to be heated to more than 161°F, the milk will be instantly pasteurized as it passes that temperature. There is no need to pasteurize the milk prior to preparing things like soup or pudding that are boiled as part of the recipe preparation.

- **Vinegar or citric acid:** Before you have cheese, you have to ripen the milk, and for some cheeses this is done with an acid rather than a culture. Vinegar, citric acid, and even lemon juice can be used. Avoid using fruit preservation products from the grocery store. Although many contain citric acid as the active ingredient, they also contain ingredients that you don't want in your cheese.

- **Cultures:** There are two types of cultures, thermophilic and mesophilic, and they are not interchangeable. Thermophilic cultures work at a higher temperature than mesophilic ones. Both thermophilic and mesophilic cultures have many varieties, and specific varieties produce specific types of cheese. You may have seen some recipes that called for commercial yogurt or buttermilk to be used as a starter culture, but commercial products may or may not contain enough live cultures to successfully make cheese, and if you wind up giving a gallon of milk to your pigs or chickens, you will not have saved any money. The recipes in this book don't call for the use of a mother culture, prepared starter, or bulk starter. It simply makes more sense for the home cheese maker to purchase freeze-dried starter cultures. Although you can freeze a mother culture, it lives for only a couple of months, and you won't know that it's dead until you've had three successive batches of cheddar threaten to explode out of your cheese press. (Do I have to tell you how I know this?)

- **Rennet:** Without rennet, cheese won't turn into curds that will melt. Cheese-making rennet is available in tablets or in liquid form. Although tablets last longer, the liquid will last for a couple of years in the refrigerator, and you can purchase as little as 2 ounces. I prefer the liquid because it is easier to measure than trying to break tablets into the correct proportions. Junket rennet in supermarkets contains about one-fifth as much actual rennet as cheese-making rennet. Although it might work in soft cheese, which needs only one drop of cheese-making rennet, it won't work in hard cheeses, and it contains a number of unpronounceable ingredients. Cheese-making rennet is available made from vegetable or animal sources, and as a genetically modified version. The strength of rennet, single, double, or triple strength, is indicated on the product label.

- **Salt:** One of my pet peeves about commercial cheese is that much of it is too salty for my taste, so when we started making our own cheese, we used little or no salt. Although this will work in fresh cheeses, you need the salt in aged cheese because the salt works as a preservative. You need a salt that has no additives, including iodine or unpronounceable chemicals. Salt labeled for canning, pickling, or cheese making is usually free of additives, but check the label to be sure.

- **Mold:** Cheeses such as brie, Camembert, blue cheese, and Muenster are ripened by mold, requiring the addition of mold to the cheese-making process. Some cheese-making recipes call for the mold to be sprayed on the cheese with an atomizer. But it makes more sense for the home cheese maker to add the mold to the milk mixture, so you don't need to purchase a spray bottle and mix up a large amount of mold and water that you won't be using. Mold is available from the same cheese-making supply places that sell cultures and rennet.

- **Calcium chloride:** Many recipes assume you are using store-bought milk, so calcium chloride is given as a necessary ingredient. When you have your own goats, you do not need to use calcium chloride. I'm listing it here because if you have other cheese-making books, you may have seen it listed as an ingredient. The recipes in this book assume you are using milk fresh from your goats.

- **Lipase:** This is another ingredient that you will not find in the recipes in this book. Lipase, an enzyme, is present in raw milk and is destroyed by high-heat pasteurization. It is not necessary in most cheeses, but it is critical to the flavor of many Italian cheeses, such as provolone.

CHAPTER 14

Dairy Products

✧ ✧ ✧

THERE ISN'T ANYTHING easier to make than fermented dairy products, and I doubt there is anything that is both tastier and healthier. You don't need any special equipment, and if you have your own milk, you only need to buy the cultures. You can make buttermilk, sour cream, and yogurt by adding the appropriate culture and letting the milk set at the right temperature for a few hours. Yes, it really is that simple! You can even replace some of the other dairy products that you may have been buying, such as coffee creamer and caramel sauce.

Buttermilk and Sour Cream

Buttermilk and sour cream are made the same way. The same culture is used, but it is added to milk for buttermilk and to cream for sour cream. Buttermilk is the easiest dairy product to make because you simply add culture to the milk, let it sit at room temperature until it has thickened, and then store it in the refrigerator. Sour cream is only slightly more work because you have to separate the cream first. Because goat milk does not separate quickly, this is most easily accomplished by using a cream separator.

The milk or cream ferments at room temperature, which means you are using a mesophilic culture. Cheese-making supply companies identify which cultures work well for sour cream and for buttermilk.

Homemade buttermilk can be used as a culture for making sour cream, but using store-bought buttermilk as a culture may yield disappointing results because it may not contain live cultures.

Although buttermilk was originally the milk left after making butter, today's commercial buttermilk is a concoction of milk and chemicals meant to resemble the original. It isn't necessary to make butter before making buttermilk. Either skimmed or whole milk can be used, but the latter will be much thicker. In fact, some people say it reminds them of sour cream, especially when it is made with Nigerian Dwarf milk, which is especially high in butterfat and milk solids.

Buttermilk can be recultured if it is relatively fresh, say no more than a week old. To reculture, we empty out the old jar of buttermilk, but we don't rinse it. Then we refill the jar with fresh milk, swirl it to mix the fresh milk with the buttermilk left in the jar, and then pour it into a clean jar.

Yogurt

The process for making yogurt is very similar to the process for making buttermilk, but it uses a thermophilic culture, which needs heat to work. The milk must be at 110°F to 120°F in order to culture. There are a number of different yogurt cultures available, including *Streptococcus thermophilus, Lactobacillus delbrueckii* subsp. *bulgaricus, Lactobacillus acidophilus.* Different cultures can create thicker, tangier, or sweeter yogurt, and you can experiment to find one that you especially like. Directions usually come with the yogurt cultures along with recommended times for culturing, but you can decide how long to culture it based on your preferences. As it cultures, the yogurt will become thicker and tangier.

A commercial yogurt maker will keep the milk at the correct temperature while it cultures. But you can also put it into a thermos or set the jar of milk with added culture in an insulated cooler that is filled with 120°F water or that has a heating pad in it that will heat the milk to a temperature in the range of 110°F to 120°F. Be sure to check the temperature of your heating pad first by putting a jar of water in the cooler to see whether the low, medium, or high setting gets the water to the correct temperature range.

Fruit-flavored yogurt is easily made by adding a tablespoon of your own homemade jam to a bowl of yogurt before serving. I also love adding maple syrup or granola to homemade yogurt.

You can reculture yogurt, essentially using it as a mother culture, provided the yogurt is not more than a few days old. After a few days, yogurt starts to loose its oomph and won't make yogurt that is thick and creamy. To reculture, we empty out the yogurt jar without scraping or rinsing and refill it with pasteurized milk. We pasteurize our milk by heating it to 170°F. We let the milk cool to 120°F before adding it to the "dirty" yogurt jar. We stir it up and then pour it into a clean jar to culture. We don't continue using the same jar because mold will start to grow around the rim after a couple of weeks.

We made yogurt with raw milk for a year or two, but we found it impossible to get consistently good results when reculturing yogurt with raw milk, which is why we started pasteurizing milk to make yogurt. Others have found that raw milk makes a thicker yogurt than pasteurized, so you will have to see which one you prefer with your milk. Because the natural bacteria in the environment vary from farm to farm, the yogurt made with raw milk will be unique to your farm. And this makes local food unique and interesting.

Sweets

In addition to all of the fermented dairy products you can make, you can also make all of the other dairy products in your life, including coffee creamer and caramel sauce.

❖ Caramel Coffee Creamer

I turned into a coffee drinker in my thirties when I discovered flavored coffee creamers. Then I read the labels and lived in a state of cognitive dissonance—my taste buds wanted the creamer, but my brain did not want the artificial ingredients. If you love flavored coffee, the good news is that it is incredibly easy to make your own caramel coffee creamer. This also makes a delicious cup of chai. I like to make this in the spring when we have lots of extra milk, and I don't mind the cooking time, because it is still cold outside. This is also a great use for skimmed milk left after separating the cream to make butter or sour cream. The skimmed milk version has fewer calories, and I don't notice a difference in taste.

✤ ✤ ✤

2 quarts goat milk, whole or skimmed

2 cups sugar

½ teaspoon baking soda

Makes 1 quart.

Put all the ingredients into a pot that is at least a gallon in size and stir over low heat to dissolve the sugar and baking soda. A large pot will contain the foam when the milk starts to boil. Yes, you really do need the baking soda. If you leave it out, the milk will boil over. (Do I have to tell you how I know this?) Be sure to put the pot over low heat on the smallest burner so the milk doesn't boil over. Once you've made this creamer successfully on your own stove, you'll know which pot to use and on which burner, and it will be a breeze. Check on the milk every hour and stir, continuing to let it simmer on low heat. As the sugars caramelize, the milk will turn tan and then darker. I like it when it is reduced by about 50 percent, but, really, you can decide it is "done" at whatever point it suits your taste. If you want it slightly less sweet, reduce the simmering time. The creamer can be stored in the refrigerator for a couple of weeks.

❖ Caramel Sauce

2 quarts goat milk, whole or skimmed

2 cups sugar

½ teaspoon baking soda

Makes 3 cups.

Follow the directions for making Caramel Coffee Creamer and continue to cook until the sauce is the consistency of a runny pudding. It will get thicker when it cools. This caramel sauce is perfect for serving over ice cream or for dipping apples. This can be stored for a few weeks in the refrigerator or longer in the freezer.

Because it takes no more effort to make this in larger batches, I often double this and start with a gallon of milk. When it is reduced by half, I pour off a quart to use as coffee creamer, and I continue to simmer the remainder until it becomes caramel sauce.

CHAPTER 15

Acid-Ripened Cheeses

✥   ✥   ✥

MANY PEOPLE BEGIN making cheese with acid-ripened cheeses because these cheeses don't require special ingredients. They are ripened with the addition of an acid such as vinegar, citric acid, or lemon or lime juice. These cheeses don't require a mold, and most are drained in cheesecloth.

Vinegar

Queso blanco or paneer is often the first cheese that a new cheese maker makes because they have all the ingredients and equipment on hand. The two cheeses are very similar to each other, except that queso blanco is made with vinegar and paneer is made with lemon or lime juice. Neither cheese melts.

✤ Queso Blanco

Although queso blanco is the first cheese that many people make, most novice cheese makers don't know what to do with it and don't make it more than a time or two. My family has made queso blanco hundreds of times, and it was the cheese that I taught my children to make many years ago. We often cube it, stir-fry in sunflower oil until golden brown and crunchy, and then add it to a marinara sauce and serve it over pasta.

✣ ✣ ✣

1 gallon goat milk

¼ cup white wine vinegar

Makes 1 pound.

Put the milk into a 5-quart pot and heat it until it reaches 190°F. Add the vinegar and stir until the curds and whey separate. The whey will be clear with a slight greenish tint to it. Pour the curds and whey into a cheesecloth-lined colander and drain over a pot to catch the whey. Tie up the cheesecloth and hang it to drain for a couple of hours. You can hang the cheesecloth on the kitchen faucet, from a cabinet knob, or from a skewer that spans the top of a large pot. This cheese can be eaten after draining, or it can be chilled for a few hours to make it easier to slice. This cheese should be used within a few days for the best flavor, but it can be stored in the refrigerator for about a week before it starts to mold and taste sour.

I like to put my cheesecloth lined colander in a milk bucket to catch the whey.

I let my queso blanco or paneer drain in a bag made by tying up the four corners of a cheesecloth. I hang the bag over a milk bucket in my sink to drain, saving the whey to feed my pigs or to make bread or gjetost.

Don't toss that whey! Whey is a nutritious liquid with protein and other minerals and vitamins. It can be used as the liquid in bread recipes. It gives a nice lift to whole grains as it is a natural dough conditioner. There are entire books written on lacto-fermentation, which utilizes whey. It is also an excellent source of nutrition for chickens, turkeys, and pigs. Chickens and turkeys drink it like they would water, and a pig will bury its mouth in it until it has slurped up every last drop. Even dogs and cats love it. If you have no other use for whey, you can use it as fertilizer in your garden or on your pasture.

Using Whey

❖ Gjetost

Gjetost (pronounced yay-toast) translates from Norwegian as "goat cheese," but it is more like fudge or caramel than cheese in both flavor and the process for making it. It is not ripened. It is simply whey that has been boiled and reduced to about 20 percent of its original volume, which makes it incredibly rich.

My first attempt to make gjetost was an icky failure. I wound up with a salty, sweet, grainy goo in the bottom of a pan after eleven hours of simmering whey on the stove. It looked like sand that had sunk to the bottom of a watered-down caramel sauce. No one else would even taste it. In spite of its dreadful appearance, it tasted delicious, and I knew I had to find a better recipe and try again. Because it is extremely rich and you shouldn't eat too much at one sitting, I make it only when we have company coming.

❖ ❖ ❖

Whey is the only ingredient you need for making gjetost, and you can use whatever amount you have left over after making cheese. You should use fresh whey, however, or the final product is likely to be too salty. Put the whey on the stove in a heavy-bottomed pot and boil. And boil. And boil. This process will recall making maple syrup for those of you with experience boiling up sap. You will be boiling for a few hours. The more surface area in your pot, the faster the whey will boil down, so given a choice between a tall, skinny pot and a wider, more shallow pot, choose the wider pot to boil down the whey faster.

When the whey starts to thicken and is as thick as a runny pudding, I use my stick blender to get rid of the graininess. When the whey is smooth, I place the pot in a sink of cold water to chill it rapidly while I whisk it.

When the gjetost gets thick enough to stay on one side of the pot when scraped over there, I know it's time to pour it into a pan for chilling in the refrigerator for a few hours before serving.

After gjetost has been refrigerated, it will become a semi-solid piece of cheese, which can be sliced for serving.

Ricotta

Although ricotta can be made from whey, the yield is so low that it is hardly worth the effort. I'd rather use whole milk and get enough cheese to make lasagna, stuffed shells, or manicotti.

✣ ✣ ✣

Heat the milk to 190°F, add the vinegar, and stir until the curds and whey separate. Remove the milk from the heat and let it sit for 10 to 20 minutes so that it can begin to cool, and then put the pot in a sink filled with cold water to the same level as the curds and whey on the inside of the pot. Stir the curds until the temperature reduces to 90°F, and then drain the curds and whey in a cheesecloth-lined colander. Hang the tied cheesecloth for about 10 minutes. Remove the cheese from the cheesecloth and put into a bowl to use immediately in your favorite recipe, or place it in a covered container to store in the refrigerator for five or six days.

2 quarts goat milk

2–3 tablespoons wine vinegar (white or red)

Makes ½ pound.

Citric Acid

Citric acid in powder form or in a citrus juice, such as lemon or lime, can also be used to culture milk. Lemon or lime juice is used to make paneer, whereas powdered citric acid is used to make mozzarella. Although you may recall using a citric acid preparation in canning, it may have included other undesirable additives, so be sure to check the ingredient list before using it for cheese making. You can buy pure citric acid from the cheese-making supply companies. To make the purest paneer, use juice from real lemons or lime, rather than bottles, which may also contain preservatives and other ingredients.

Paneer

Palak paneer and saag paneer, which are spinach dishes, and mattar paneer, which uses peas, are popular dishes in Indian cuisine. Making your own paneer is just as easy as making queso blanco.

✣ ✣ ✣

2 quarts goat milk

2–3 tablespoons lemon or lime juice

Makes ½ pound.

Many traditional paneer recipes suggest heating the milk until it boils and then adding the lemon or lime juice. This method will work well, but don't leave the kitchen while heating the milk. If you've ever had milk boil over, you'll know why I'm saying this. If you prefer to use a thermometer, gently heat the milk to 190°F and then add the juice. Stir the milk until the curds and whey separate. Drain the curds in a cheesecloth-lined colander. After hanging to drain for about 5 minutes, place the cheesecloth-wrapped curds in the colander again and put weight on it to force out a little more whey. A cast iron skillet or a pot with 2 cups of water in it will work for the weight. Remove the cheese from the cheesecloth and store it in the refrigerator. Like queso blanco, paneer has no culture to preserve it, so it must be eaten within a week.

❖ Mozzarella

1 gallon goat milk

½ tablespoon citric acid

¼ teaspoon liquid rennet

Makes 1 to 1¼ pounds, depending on the butterfat content.

Mozzarella can be made with acid or with a culture. Acid-ripened mozzarella is meant to be eaten within a few days of making it, while culture-ripened can be kept up to three or four weeks. The culture-ripened cheese takes much longer to make, however. Because we eat a pound or two of mozzarella every week, we don't have time to make the culture-ripened version. We make this recipe so often that it takes us only about twenty minutes from start to finish. This is the most forgiving cheese I've ever made. Over the years, we have done just about everything "wrong" at one time or another, and it still turns into mozzarella in the end.

✧ ✧ ✧

If the milk is refrigerated, bring it up to 55°F over low heat and then sprinkle the citric acid into the milk while stirring constantly but gently. If the milk is fresh from the goats, strain the milk into a pot, and when the milk has cooled off to below 90°F, add the citric acid. Heat the milk to 90°F, and add the rennet while stirring constantly. Without turning off the heat, continue to stir gently as the curds and whey separate. The temperature will continue to rise. Stop stirring when the whey is clear with a slight greenish tint, turn off the heat, and let the curds sit for 15 minutes to sink to the bottom of the pot and "knit," which means the curds will stick together in one large mass.

Microwave directions: Using a large slotted spoon, remove the curds and put them in a microwaveable bowl. Heat the curds for 1 minute on high. Wearing thick rubber gloves, remove the curd mass from the bowl and start to knead it like bread dough. As you squeeze the curds, whey will be pushed out. You'll probably want to do this over your sink, or you'll have a big mess to clean up. When it gets difficult to squeeze the curds, put the mass back in the bowl and microwave for another 30 seconds. Repeat the kneading. If the curd mass becomes difficult to knead but you are still squeezing out a lot of whey, you might need to microwave for another 30 seconds and knead again. When most of the whey is worked out of the curds, you will be able to stretch the curds for at least a foot without breaking. Removing the whey is important because the cheese won't last more than a few days if excess whey is trapped in the curds, and it will be more difficult to shred if it is too moist.

Stovetop directions: Rather than letting the curds sit immediately following the separating of curds and whey, continue slowly heating the curds to 140°F without stirring. Once the whey is at 140°F, turn off the heat and let the curds sit for 5 minutes to knit. Press the curds against the side of the pot to form a solid mass. Then scoop out the curds with a slotted spoon and put them into a bowl. Follow the directions above for kneading. You can usually remove all of the whey without needing to reheat the curds.

Although this mozzarella cannot be aged, it can be frozen. We make extra mozzarella through the summer when we have extra milk so that we'll have plenty of mozzarella for our weekly pizzas during the winter when few of our goats are milking.

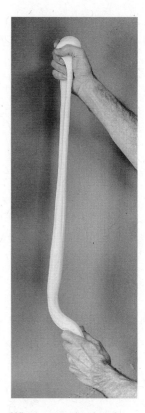

When enough whey has been kneaded out of mozzarella, you should be able to easily stretch it a couple of feet. If it breaks rather than stretches, there is still too much whey in it, so you should continue kneading a bit longer.

CHAPTER 16

Culture-Ripened Cheeses

CULTURE-RIPENED CHEESE was one of the first preserved foods created by humans, and it evolved differently in response to the unique milk, environment, and cheese-making habits of different geographic areas. Even if you follow every recipe in this book exactly as it is written, your cheese will taste different from mine, perhaps subtly or maybe drastically different. Cheese making may be more chemistry than cooking, but farmstead cheese making is as much art as it is science. The big cheese factories took over cheese making in the United States in the nineteenth century and standardized cheeses to a level never before seen in the world. But variety is the spice of life, and those of us making our own cheese have the opportunity to make a product that is not only as good as a commercial cheese, but is actually better. While you may want to start out making familiar cheeses, with experience you will probably find you want to experiment, changing the recipes to suit your own taste.

One of the most important things that any cheese maker needs to do is to keep a journal. I cannot overemphasize the importance of writing down everything that you do when making cheese. Although you might be able to connect the dots when you make a fresh cheese that turns out better or worse than expected, it is impossible to remember what you did six or nine months ago when you made that cheddar that won't melt or the Gouda that tastes like cheap store-bought cheese.

This is one of those lessons that I learned the hard way. I was terrible at maintaining a journal when I started making cheese, and my aged cheeses were reliably terrible. When my husband started making aged cheeses, he kept notes that reminded me of the lab reports we had to do in science classes. His first aged cheeses were not great, but he would write down tasting notes as we opened each one over the months. Through the years, his aged cheeses have improved dramatically as he has tweaked his recipes to work with our Nigerian Dwarf milk. Having milk that is much higher in butterfat and milk solids than the average goat or cow milk taught us early that it's okay to stray from the recipe and that our cheese would be better for our experimentation.

Choosing Cultures

There are a number of different mesophilic cultures with scientific names, and they vary slightly in the way they react with milk. However, home cheese makers don't have to figure out everything on their own. Cheese-making supply companies have combined some of the most popular cultures into packets of freeze-dried starters and given them simple names like MA, MM, MD, and flora danica, and in the catalog or website description, they give recommendations for which culture to use when making specific cheeses.

When we started making culture-ripened cheeses, we purchased several types of cultures and used them to make the recommended cheese. Over time we began to experiment with using whatever culture was on hand to make our cheese recipes. Over the years, we've pretty much started using MM culture for all the cheeses we make that require a mesophilic starter, including chèvre, feta, Camembert, cheddar, Gouda, and Havarti. This is why my recipes don't specify a culture beyond stating whether it needs to be mesophilic or thermophilic. Go ahead and try using the culture specified by the supplier, but don't be afraid to try a different one. You may discover, as we have, that when combined with your unique milk, you can create a tastier cheese by using a culture that is different from what is recommended.

◈ Chèvre

This is the cheese that is often referred to as "goat cheese" in stores and on restaurant menus. And the word, which is French, translates as "goat." In addition to serving chèvre with crackers or crusty bread, we love to have it on sandwiches with grilled Portobello mushrooms or to use it as the stuffing in cheese blintzes.

Although you can buy direct set cultures to make chèvre, it is incredibly simple to make it using mesophilic culture and rennet.

✧ ✧ ✧

We make this cheese using milk that has been pasteurized because in our experience it creates a more spreadable cheese than when made with raw milk. We pasteurize the milk at the time we are going to make the cheese by heating it to 145°F and holding the temperature for 30 minutes. Then we put the pot of milk into a sink filled with cold water to reduce the temperature to 86°F.

When the temperature has reduced to 86°F, add the amount of mesophilic culture recommended by the supplier to ripen 1 gallon of milk, and stir gently for about 30 seconds to combine. Add the rennet and continue stirring for about another 30 seconds. Cover the pot and set it aside, where it will be undisturbed for 10 hours or until the curds have a clean break, which is when the curds form a solid white mass, the whey is almost clear, and the curd holds its shape when you cut into it. When you have a clean break, spoon the mass into plastic molds and let each one drain for 12 hours.

It took me years to make a good chèvre, partly because I kept no notes and changed my methods slightly from one batch to the next and partly because I had not learned enough about cheese to know what caused the variations. If your cheese turns out moister or drier than you want, you can usually fix the problem by adjusting the size of the curds or the amount of time you are waiting before putting the curds into the molds.

When chèvre is so moist that it continues to drain for days after you take it out of the molds and put it into a storage container, try reducing

1 gallon goat milk

mesophilic culture

1 drop rennet in
   2 tablespoons water

Makes 1 to 1¼ pounds, depending on butterfat content.

the size of the pieces of curd. Read the section on cutting the curds for an explanation of how curd size affects moisture. You may be waiting too long to put the curd into molds. A lot of chèvre recipes suggest the amount of time is not important, but if you let the curd continue to sit in the whey after it has reached the clean break stage, it will reabsorb the whey, making the curd weaker, and you'll have a very mushy cheese that doesn't hold its shape when removed from the mold.

Flocculation

Many cheese recipes specify a waiting time between adding rennet and cutting the curds. It is usually somewhere between 30 minutes and an hour. Typically, recipes that specify the waiting time also say "or when you get a clean break," which means when you can cut the curd cleanly and it stays cut. Cutting the curds too soon results in a weak curd that will not produce a cheese with a desired texture.

The amount of rennet used, the temperature of the milk, and the composition of the milk can all affect the flocculation rate, which is the moment at which the milk begins to gel. The point of flocculation will help you determine when to expect a clean break. This means that

You have a clean break when the curd holds its shape when you cut into it.

the optimal waiting period can be different for your milk or your goats' stage of lactation. You can figure out the ideal waiting time by checking flocculation. Using an eyedropper, put a drop of milk into water. If the milk retains its shape as a little ball, it has begun to gel. If it disperses into the water like liquid milk, wait a couple of minutes and try again. You can also check flocculation by floating a lightweight plastic container on top of the milk. If you can spin the container, the milk is liquid. If the container won't move, the milk has begun to gel.

To determine the number of minutes that should pass before cutting the curds, multiply the flocculation time by a number that will vary depending on the type of cheese you are making, as indicated below. For example, if you are making Camembert, you should see flocculation at about 10 minutes. If you do, then you multiply 10 by 5 or by 6, which means that 50 or 60 minutes should pass between the time you added the rennet and the time you cut the curds.

Cheese Flocculation Times

Cheese	Flocculation factor	Target cutting time
Alpine, Parmesan, and other extra-hard cheeses	2–2.5	20–30 minutes
Cheddar, Gouda, and other semi-hard to hard cheeses	3–3.5	30–50 minutes
Feta and blue cheese	4	50–60 minutes
Brie, Camembert, and other bloomy rind cheeses	5–6	50–60 minutes

If you find you need to cut much earlier or later than the target cutting time, something is not quite right. If you are following the temperature suggested in the recipe and flocculation is happening earlier, reduce the amount of rennet in future batches. If you are going beyond the target time by more than 10 or 15 minutes, increase the rennet slightly or try replacing the rennet with a fresh supply. Rennet can lose its potency if it is stored improperly or kept beyond the expiration date. Following the estimated times in recipes usually produces good results, but determining the optimal time to cut the curds and adjusting the amount of rennet accordingly will produce a cheese that melts better and has a smoother texture. When we learned about flocculation, we discovered that we could reduce our rennet by more than half of what we had been using! The change immediately created a cheddar that melted better.

Cutting Curds

All aged cheese recipes and even a few fresh cheese recipes tell you to cut the curds into a specific size, usually an inch or smaller. When you have a clean break, cut the curd mass from top to bottom into strips using a long knife. Turn the pot 90 degrees, and cut again to create a grid, as shown in the picture. Next, cut horizontally or diagonally to create curds of the appropriate size. You can use a knife and go at it diagonally, or you can use an egg turner and slide it into the curds horizontally.

Cutting the curds is more of an emotional challenge than a physical one because the curds won't all be the perfect size. The moisture content of the cheese is related to the size of the curds. The smaller the curd, the more surface area is exposed, allowing whey to drain. So, smaller curds equal drier cheese, and larger curds equal moister cheese. Chèvre is as moist and spreadable as it is because the curds are not cut at all; you are simply spooning them into molds.

After being cut from top to bottom, the surface of the curd mass will look like graph paper. Now it is time to make the horizontal or diagonal cuts to create curds the size required by the recipe.

❖ Feta

Most feta sold in the United States is made from cow milk, but it was traditionally made with sheep milk and sometimes blended with goat milk. Lipase appears in feta recipes that assume pasteurized cow milk is being used, but this recipe assumes you are using your own goat milk that has not been heated to more than 145°F during the pasteurization process.

1 gallon goat milk
mesophilic culture
½ teaspoon liquid rennet
cheese salt

Makes 1 to 1¼ pounds, depending on butterfat content.

✧ ✧ ✧

Add the mesophilic culture at the rate recommended by the supplier to milk that is at 86°F and stir gently. Cover the pot and let the milk ripen for 1 hour.

Traditional Curd Cutting

Jim Wallace, technical advisor at New England Cheese Supply, modified a whisk to create his own version of a traditional Italian *spino*, which is used to cut the curds. "The name is derived from an ancient tool that was simply a long stout branch with 4-inch to 6-inch stubs remaining from side branches that were cut off," says Jim. "Cutting with this is a bit of an art and hard to explain. You just have to see it work. It does a beautiful job once you gain some practice with it."

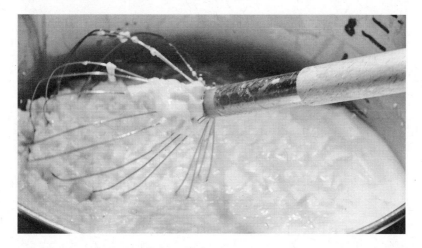

This modified whisk emulates a traditional curd-cutting tool.

Add the rennet and stir for about a minute. Check for flocculation and multiply by 4 to determine when to cut the curd. After waiting the required time, check the curd for a clean break.

Cut the curds into ½-inch cubes. After cutting, let the curds rest for a few minutes, and then begin stirring them. Keep the curds at 86°F either by placing the pot into a water bath or by turning on the burner for brief periods of time. Stir for 30 minutes.

The curds need to be drained after stirring. You can use either a cheesecloth-lined colander or plastic molds. The curds can be left to drain by gravity, or, for a slightly drier cheese, weight can be added on top of the curds to force whey out under gentle pressure.

When using two plastic molds, I set one on top of the other for 30 minutes and then switch them so that the one that was on bottom is on

When two molds of feta are draining, you can stack them on top of each other to provide a little pressure during draining that will force more whey out and result in a drier cheese.

A cheese mat helps your cheese drain by keeping the cheese from sitting in whey.

top, and I leave this for another 30 minutes. After each mold has been on the bottom, take the cheese out of the mold, flip it over, and put it back into the mold. Allow the curds to drain for 6 to 12 hours unstacked, during which time they will have matted together into a solid piece of cheese.

Cut each cheese round into quarters. If the cheese is to be eaten within a week, salt all sides liberally and store the cheese in the refrigerator.

If you are planning to age your feta, remove it from the molds, liberally salt all sides and place the cheese on a cheese mat to continue draining. The cheese should dry at 50°F to 55°F for a couple of days before brining. If you don't have a cheese cave for aging, you can put it into the refrigerator at this point. At refrigerator temperatures cheese does not really age, meaning that it does not develop a rich flavor.

Brining will preserve cheese and keep it from molding. Mold starts to grow on salted feta after a couple of weeks. To brine the cheese, put the quarters into a brine solution of 2–3 ounces of salt to a quart of water. Use a container that is filled to the very top to avoid mold growth on the top of the cheese. The cheese can be stored in the solution for months.

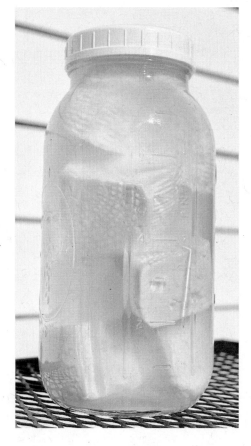

A wide-mouth, half-gallon canning jar is one option for storing feta in brine because the opening is large enough for the wedges of cheese to fit.

◆ Camembert

This bloomy rind cheese is an easy one to make. Traditionally a mold that is open on both ends is used, but a plastic basket mold can be used.

Pasteurized milk is preferred for this cheese because the cheese is usually aged only a month before being eaten. The short aging period means this cheese is more likely than most aged cheeses to develop bad bacteria because it has not aged long enough for the good bacteria to beat out the bad guys, if there are any in the milk.

✧   ✧   ✧

1 gallon pasteurized goat milk

mesophilic culture

Penicillium candidum

Geotrichum candidum

⅛ teaspoon liquid rennet

cheese salt

Makes 1 to 1¼ pound, depending on butterfat content.

Heat the milk to 90°F and add the amount of culture and mold specified by the manufacturer. Stir well after each addition, and let the milk ripen for 30 minutes. Add the rennet and check for flocculation after 5 minutes. It should happen after 5 to 10 minutes, which means a clean break will occur after 50 to 60 minutes. Cut the curds into 1-inch pieces, and let the cubes rest for 5 minutes. Then stir the cubes 5 minutes to release a little of the whey.

Spoon the curds into two or three molds for draining. If the butterfat is not terribly high, use two molds. If you have Nigerian Dwarves or Nubians, or if it is fall, when goat milk tends to have a higher butterfat content, use three molds. Do not overfill the molds. Mold-ripened cheeses need to be fairly thin after draining so they can develop a creamy interior throughout.

After 1 hour, remove the cheese from the mold carefully and flip it over. Flip it again at 1-hour intervals until the cheese has rested on each side twice. Let the cheese drain for 24 hours, and then sprinkle ½ teaspoon salt on each side.

This cheese should be aged at a temperature between 50°F and 55°F with very high humidity. This cheese needs to be isolated from

The first time I made a mold-ripened cheese I made two mistakes in aging. First, I put the cheese on a cookie cooling rack rather than on a cheese mat, and second, I didn't turn the cheese regularly. Consequently, the soft cheese oozed between the wires and the mold grew around the cooling rack.

Camembert is usually considered ready when the interior is creamy. You will know this by feel when you are flipping the cheese daily. The cheese initially feels quite firm, but after three to four weeks, you will notice it softening to the touch. If when you cut it open the texture is not ideal, note whether you would like it softer or firmer, and next time adjust the aging time accordingly.

other cheeses while being aged so that the *candidum* molds don't migrate and mold-ripen everything nearby. Storing the cheese in a plastic container solves the problem. Place the cheese on a cheese mat in the container so that the cheese won't be sitting in water, and turn the cheese daily. If you forget to turn it, the mold could wrap around the cheese mat, and separating the two would require ripping the mold off the cheese, effectively ending the aging process. Within a few days, white mold is visible growing on the surface of the cheese, and by two weeks the cheese will have a nice fuzzy coat. Unlike a true aged cheese, Camembert does not continue to improve with age and is usually best eaten around three to five weeks after making it.

Semi-Hard and Hard Cheese

Once you have perfected a variety of soft cheeses, you might want to try a hard cheese, such as cheddar. Hard, aged cheeses are made using a cheese press, which forces moisture out of the cheese. Recipes that use a press will tell you how long to press the cheese and at what pressure, measured in pounds.

Line the press with cheesecloth before putting the curds in it. The lining serves two purposes. First, it keeps the curds from sticking to the press, and second, it wicks away moisture and helps the cheese drain.

After putting the curds into the press, fold the cheesecloth to cover the top of the curds. Then place the follower on the covered curds, and press. After the initial pressing, remove the mass of curds from the press, remove the cheesecloth, flip the cheese over, and put it back into the cheesecloth-lined press.

This time, don't fold the cheesecloth lining over the top of the cheese. Instead, use a new piece of cheesecloth that has been cut to fit on top of the curds. Folding the cheesecloth over the top of the cheese results in pockets and divots in the cheese, and you want a smooth finish. You can leave the ends of the cheesecloth hanging over the edge of the press during pressing.

Most recipes recommend pressing several times, flipping the cheese over between each pressing, and using a heavier weight with each successive pressing.

❖ Antiquity Oaks Heritage Cheddar

This is the recipe that we use most often for the cheese that we call cheddar because it tastes like cheddar when made with our Nigerian Dwarf goat milk. Unlike an authentic cheddar, this recipe omits the cheddaring step, and unlike most cheddars, it is a washed curd. It all started with a Colby cheese that we aged far longer than a Colby is normally aged. We took it to a party and wouldn't tell anyone what kind of cheese it was. Everyone said the cheese was an aged cheddar. Over the past few years, we have changed a number of things about the cheese to suit our tastes. In early fall, as the butterfat increases in the milk, we gradually decrease the amount of milk and rennet we use in the recipe, and by mid winter, we are using only 1½ gallons of milk and ¼ teaspoon of rennet, and we are still getting a 2-pound round of cheese.

❖ ❖ ❖

Heat the milk to 86°F and add the mesophilic culture. Let the milk sit for 1 hour. Add the diluted rennet to the milk. Stir for 1 minute, and then let the milk rest for up to 30 minutes. Check for flocculation after 5 minutes. When flocculation occurs, multiply by a flocculation factor of 3 to calculate the resting time for the cheese to reach a clean break. When the cheese has reached a clean break, cut the curds into ½-inch pieces. Let the curds rest for 5 minutes, and then gently stir them for 5 minutes. Continue stirring every few minutes to keep the curds from matting together while you raise the temperature 2°F every 5 minutes to 102°F. Maintain the temperature at 102°F for 30 minutes and continue to stir occasionally.

To "wash" the curds, pour off the whey to the level of the curds and add 1½ to 2 quarts of 60°F water to the pot. The goal is to reduce the temperature of the curds to 80°F. Maintain the temperature at 80°F for 15 minutes while continuing to stir occasionally. Drain the curds and whey through a colander. Let the curds sit in the colander, draining, for 20 minutes. The curds will have matted by this time. Gently break them apart and sprinkle 2 tablespoons of salt into the curds and mix.

Put the curds into a cheese press and press at 20 pounds for 20 minutes. Keep an eye on the pressure and adjust the pressure gauge as needed to maintain 20 pounds of pressure because the cheese will compress considerably in the first few minutes. Remove the cheese, flip it, rewrap it, put it back into the press, and press at 30 pounds for 20 minutes. Remove the cheese, flip, rewrap, and press at 40 pounds for 1 hour. And for the final press, remove the cheese, flip, rewrap, and press at 50 pounds for 12 hours.

Remove the cheese from the press and air dry in the cheese cave for several days. Seal the cheese in plastic or wax it and store it for at least 60 days if you used raw milk. Although you can eat it sooner if you use pasteurized milk, the flavor improves dramatically with time. We've aged many blocks of this cheese for well over a year and have even aged some for more than two years. If you plan to age the cheese for six months or longer, use cheese wax rather than sealing in plastic. Age the cheese between 50°F and 55°F.

2 gallons goat milk

¼ teaspoon MM100 mesophilic culture

½ teaspoon liquid rennet diluted in ¼ cup water

cheese salt

Makes 2 pounds.

You can use a natural bristle brush to apply cheese wax to cheese for aging. The temperature of the wax should be about 220°F to kill any mold spores on the surface of the cheese. Be careful not to overheat the wax because it is flammable. Reduce the heat before applying the second coat of wax.

✤ Traditional Cheddar

After years of tweaking our own version of a cheese that tastes likes cheddar, we attended a cheese-making workshop with Jim Wallace, technical advisor for New England Cheesemaking Supply. Jim's original recipe uses 6 gallons of milk, but we reduced it to 2 gallons for a more manageable size, both in terms of milk needed and final weight of cheese.

✤ ✤ ✤

2 gallons goat milk
mesophilic culture
½ teaspoon liquid rennet
cheese salt

Makes 2 pounds.

Heat the milk to 86°F and add the culture. Stir for 1 minute, and let the milk ripen for 1 hour. Add the rennet and stir for another minute. The target flocculation is 18 minutes. Multiply by a factor of 2.5 to determine the waiting time before cutting the curds. After achieving a clean break, cut the curds into ½-inch pieces. Stir the curds gently for 15 minutes, and then raise the temperature 2 degrees every 5 minutes to 102°F. Hold the temperature at 102°F and continue stirring for 30 minutes. Pour the curds into a cheesecloth-lined colander and wrap the curds tightly in the cheesecloth.

To "cheddar," put the colander into an insulated picnic cooler with three 1-quart bottles of hot tap water to maintain a warm temperature in the cooler. Place a small cutting board across the top of the wrapped curds in the colander, and place a 1-quart bottle of hot water on top of the cutting board as a weight. After 30 minutes, flip the curd mass, and add another bottle on top of the cutting board to increase the weight. After another 30 minutes, flip the curd mass again and return the two bottles on top as weight.

After 90 minutes in the cooler, the curds will have formed into a solid mass. Weigh the curd mass, and then tear it into ½-inch pieces. Gradually add an amount of salt calculated at 2 percent of the weight of the curd mass, and mix it into the curds. Put the curds into a cheesecloth-lined press and press at 35 pounds for 1 hour. Remove, rewrap, flip, and press at 50 pounds for 2 hours. And finally, remove, rewrap, flip, and press at 50 pounds for 24 hours.

Place the cheese on a cheese mat to dry in the cave for a few days, and then add two coats of wax. The cheese should be stored at a temperature in the range of 50°F to 55°F. As with other cheddars, the flavor should improve with age.

CHAPTER 17

Meat

✦   ✦   ✦

IT WASN'T THAT LONG AGO that I said I could never eat goat meat. My goats were like pets to me. Each one had a name and a personality, just like my cats and dogs do. And then there was Hercules. He was a LaMancha buck that was especially skilled at getting over fences. He could go over a livestock panel, as well as a 5-foot-high wooden fence. Twice I caught him in a pasture with my Nigerian Dwarf does.

After he slipped into the doe pasture the first time, I moved him to a paddock almost a quarter of a mile from the does, but he jumped the fence and beat me back to the barn. After the second time he joined the does, I decided he had to go, so I posted a for sale advertisement online and in the state dairy goat newsletter. No takers. Finally, I decided that I would castrate him and train him as a cart wether so that he could help us haul firewood and plow the garden and things like that. Unfortunately, even after he had no hormonal motivation to join the does, he still wanted to go wherever his heart desired. He began crushing the woven wire fences all around the farm, and when we tried to string electric wire across the top, he didn't care. He then started to teach his bad habits to my two Nigerian Dwarf wethers.

I was worried that he would damage the fence and that coyotes would get into the pasture before we realized there was a breach. I was also worried that the wethers would damage the fence letting all of the does out and exposing them to all sorts of predators as well as to cars. Eventually, I saw no option other than to butcher the three wethers, so we took them to be processed professionally. I quickly learned that although old goat may not taste too bad, it causes terrible indigestion. We tried many recipes, and everyone always ended up with indigestion for several hours after eating the meat. We finally gave up and fed the rest of the meat to our livestock guardian dogs.

As it turned out, that experience was my introduction to raising goats as meat animals. The next year we were blessed with twenty-nine bucklings. One of the things I had always liked about Nigerian Dwarves was how easy it was to sell them as pets, pasture ornaments, or weed eaters. However, in most years we would have only a dozen or so bucklings. Twenty of the bucklings eventually sold, leaving us with nine as we headed into fall. We made a plan to butcher them ourselves and eat them. My husband butchered one in October, which we ate, but we never got around to butchering more. When spring came, I decided to offer them at a discount to people who were looking for pets, but everyone wanted babies, even if they cost more than the yearlings.

As summer came, we headed into a drought. I had always believed that none of our animals would ever go hungry because we had thirty-two acres of pasture and woods and I was careful to not overstock. But after a couple of months with no rain, there was no grass left. I called my usual hay suppliers, even though I didn't expect good news. No one near me had any hay. I eventually wound up buying some from out of state. The forecast for the drought was not good, and I realized we had eight wethers that were not likely to ever sell as pets, so we decided to butcher them as well as three mature bucks that we really didn't need as breeding animals. Although I originally didn't plan to eat our goats, I have come to realize that just as the goat milk nourishes us, the meat can too.

Meat Quality

Goat meat, called chevon in France or cabrito in Mexico when talking about meat from a kid, is lower in fat than beef, pork, lamb, or chicken, making it very healthy. Meat from younger goats is preferred, and because a doeling has a future as a milker, we are usually talking about a buckling being used for meat. It is generally thought that the best time to butcher wethers is from four to twelve months of age, but there is disagreement when talking about bucks. As intact bucks get older, they are more likely to start peeing on themselves and developing a bucky odor, which means the meat could have an unpleasant flavor or aroma. However, intact bucks grow considerably faster than wethers, which means there is more meat to be had from a buck at a younger age. On the other hand, if a castrated kid is able to nurse longer, the weight gain from nursing may compensate for the weight loss usually associated with castration. We castrated a Nigerian Dwarf buck and let him continue to nurse until we butchered him at eight months. His live weight was 51 pounds, which is excellent for a Nigerian buck of that age.

Of course you can butcher animals of any age, but the quality of the meat changes. My local meat processor says that any meat from an animal over three years of age will not be appetizing to those of us who didn't grow up eating older goats.

Butchering

If you are new to a homesteading lifestyle, you may have the same reluctance to butcher your animals that I did when we got started. But when you breed goats every year to produce milk, you have to do something with the kids. This doesn't mean you have to butcher them yourself or eat them. You can avoid butchering by selling the extra wethers or bucks live for slaughter. The demand for goat meat in North America exceeds supply, and each year goat meat is imported.

Commercial services

There are alternatives to butchering goats yourself. A local custom meat processing plant will butcher animals for you. This involves transporting the animals to the plant, which can be stressful for them, so it is best to take more than one. Just as goats are happier living with friends, they are also happier going on a trip with a friend or two. And the quality of the meat is said to be better when the animals are not stressed at slaughter time.

Mobile butchering services are ideal because the goats do not need to leave home, which reduces their stress even more. A mobile meat processor is a USDA-inspected facility that is housed in a semi trailer, which brings the slaughterhouse to your farm. Unfortunately, these are a fairly new development and not available in all areas, but their popularity is growing.

How Much Meat to Expect	About 50 percent of the goat's live weight remains after butchering and removing all the non-edible parts. This is known as "hanging weight," and it is not all meat. When we butchered the 51-pound Nigerian wether, we decided to debone it. The result was 14 pounds of bones and 14 pounds of meat. A standard breed goat will probably have a live weight of 80–100 pounds before a buck or wether reaches a year of age, which would yield a 40–50 pound hanging weight.

Home butchering

Some people prefer butchering at home because the animals are never stressed by being taken somewhere strange at the end of their life. The commonly used methods of slaughtering animals at home are to shoot a goat in the head or to cut the jugular vein in the throat. The recommended approach for shooting an animal is to shoot at the center of an imaginary X between the goat's eyes and ears. After shooting, the goat will fall down and its body will jerk. This movement is natural. An animal shot in the pasture as it is grazing will have no idea that anything is amiss.

Cutting the jugular vein and "bleeding out" an animal is the other method of slaughter commonly used in home butchering, and some religions require this method to be used. Some people say it is less traumatic for the goat than shooting.

On their farm in Ohio, Julie Gorrell's husband slaughters Kosher. The knife blade is sharp and free of any nicks. "In one smooth stroke, he slits from one ear to the other," she explains. "He usually does it while they are calmly grazing beside him. He always likes to keep the animal calm and feeling cared for until their last breath. No 'stringing them up by the feet' while they are alive like I've heard some people talk about who want to buy goats to slaughter. We don't sell to those people."

Learning to butcher is easier if you start with a smaller animal because it is simply easier to handle, and you don't need a large work surface. You might want to limit your first attempt to a goat that is small enough to cut up on your kitchen counter. As with many of the other aspects of raising goats, it is a good idea to have an experienced person help you the first time you butcher a goat. Experience butchering deer or sheep is helpful because the anatomy of these animals is similar to that of goats. The first few times my husband took animals to our local processing facility, he watched them kill, remove the head and skin, and eviscerate. At that point, the carcass is put into a cooler for three or four days before they cut it up. Because goats have very little fat on them, they do not benefit from long aging.

Don't get caught up in creating perfect cuts of meat when you are butchering. If you don't have a bone saw and other butchering equipment, it is entirely acceptable after skinning a goat simply to cut all the

---

Leather

If we are going to take the life of one of our animals, I feel that no part of the animal should go to waste. In addition to feeding the raw bones of our butchered goats to our livestock guardian dogs, we keep the skins to be tanned and used as leather. You can purchase home tanning kits to tan the skins. But if you are not quite ready to learn butchering or tanning, you can ask your processor to save the skins for you, and you can send them to a tannery. Skins need to be dried before being sent to a tannery. After skinning the hide off the animal, the fat should be scraped off the inside of the hide using something blunt like the side of a spoon. Using a knife can result in accidentally cutting through the hide. Once you have removed the fat, salt the hide heavily. My rule of thumb is that if flies are attracted to the hide, I haven't used enough salt. Within a couple of weeks, the hide should be dry enough to ship to a tannery. Depending on what you want to make with the leather, the hide can be tanned with the hair on or the hair removed.

---

meat off the bones to create fillets, small boneless roasts, stew meat, and ground meat. The meat that runs along the spine is the most tender and makes delicious little medallions for stir-frying. You can give the raw bones to your livestock guardian dog. In fact, raw bones are safer for your dog than cooked ones, which can splinter and cause intestinal damage.

After the hide is dried but before it is tanned, it is rawhide, and although it can bend, folding it could result in cracking. Keep this in mind when packaging to ship to a tannery. The tannery can split a thick hide from an older animal into top grain and suede for garments. Leather from a young goat is very thin and works well for garments. Few people today realize that "kid gloves" refers to gloves that are made from the skin of a young goat. Thicker skin from an older goat could be used to make moccasins. Leatherworking equipment and patterns for making leather accessories can be found online.

Cooking

If you butcher at home, it is unlikely that you have a commercial cooler to hang an entire carcass to age before cutting it up. You can, however, cut up the meat and store in your refrigerator for a few days before

---

❖ MARGARET LANGLEY, Mobile, Alabama

I am totally in favor of home slaughter. Less stress, cheaper, cleaner. Remember, if your meat is cleaned elsewhere, you are relying on someone else to keep your meat from being contaminated by the built-up bacteria of a slaughterhouse and any disease that may be present from all the other animals that have gone before your animals. My husband was a truck driver and hauled cattle on a large scale, so we know how many cattle that are sick get slaughtered, and we do not want our meat cleaned in a place like that.

We killed two bucks this year. They were walked calmly out behind a building far away from the herd (we never do this in sight of the herd!), told bye and petted and hugged. They had no fear that anything was wrong. We respect their lives and appreciate the sacrifice of each one to provide for our family. A sharp knife to the jugular does seem best to us. It is very Biblical, quick, and clean. They are then hung, drained and cleaned like deer. Pelts are stretched out, scraped, salted and saved, not wasted. They are beautiful and serve to remind us of the good times we shared with animals that we loved and

After a week or two, the skin will be completely dry and inflexible. It is now ready to be tanned.

cooking. Like most meats, the meat from a younger goat has a milder flavor, will be tenderer, and will be able to handle higher heat cooking methods, such as grilling. Meat from an older animal will need to be marinated and cooked more slowly over a lower heat. Keep in mind that

respected and of the sacrifice made by our animals to provide for our needs. Any death should be sad, but the death of an animal you know seems sadder than the death of one you've never seen. We really should feel that sad about all the animals we consume, but it just doesn't seem so bad when we pick up those packs of processed meat at the store.

I guess what I am trying to say is that slaughtering an animal should be a hard thing to do because that means we care, but if we are going to eat meat, it should also be something we are willing to do. Sometimes we benefit most from the things that are the hardest. Slaughtering is difficult, but it will provide our families with the healthiest meat we could possibly give them and I promise you the best tasting meat you have ever had.

goat meat is the most low-fat meat available, which means it is easy to overcook it and dry it out.

Of course, you can substitute goat meat for beef or lamb in any of your favorite recipes, but the following recipes complement goat meat's unique flavor.

❖ Indian Goat and Sweet Potatoes

1 pound ground goat meat

2 pounds sweet potatoes, peeled

1 onion, chopped

1 tablespoon oil

½ teaspoon salt

1 teaspoon ground coriander

½ teaspoon garam masala

1 cup goat-milk yogurt

2 tablespoons lemon juice

1 clove garlic, crushed

Serves 4.

Chop the sweet potatoes into 1-inch cubes. Put them into a 2-quart pot and cover them with water. Boil the sweet potatoes for 20 minutes or until a fork inserted into a cube breaks it in half easily.

Chop the onion and begin browning it in oil. Add the ground goat, salt, coriander, and garam masala. Stir the meat frequently to prevent burning until it is cooked thoroughly. Add the cooked sweet potatoes to the pan, and stir the cubes into the meat mixture.

In a bowl, mix together the yogurt, lemon juice, and garlic.

To serve, put the meat and sweet potato mixture on a plate and drizzle it with the yogurt sauce.

❖ Goat Goulash

2 pounds goat stew meat

3 garlic cloves, crushed

1 tablespoon paprika

¼ teaspoon cayenne pepper

½ teaspoon salt

3 tablespoons oil

1 onion, sliced

1 pound carrots, sliced

2 tablespoons unbleached flour

1 cup goat milk

1 cup water

Serves 8.

Preheat the oven to 325°F.

Put 1 tablespoon of oil in a 5-quart cast iron Dutch oven. Add the stew meat and set on medium heat to brown. Add garlic cloves, paprika, cayenne, and salt to the meat and stir. After 5 minutes, add the onion and carrots to the pan, and continue stirring to prevent burning. Cook for 10 minutes and then remove the pan from the heat.

In a small skillet heat 2 tablespoons of oil, and whisk in 2 tablespoons flour to make a roux. Add the goat milk and water, continuing to whisk while bringing it to a boil. Pour the sauce into the meat and carrot mixture. Cover and bake at 325°F for 1 hour or until the meat is tender.

CHAPTER 18

Soap

✦ ✦ ✦

Shortly after becoming a goat owner, I decided I wanted to make goat milk soap, but all the talk of lye blinding you or burning holes in your skin scared me off for a few months. Then I met a soap maker who showed me how to make soap. After seeing the process in person, I purchased the ingredients and equipment and made my first batch of soap with water as the liquid rather than goat milk. I was milking only one goat at the time and I didn't want to waste a drop of milk on a batch of soap that didn't turn out as it was supposed to. After making two successful batches with water, I switched to goat milk and haven't made a batch with water since. I can't go back to store-bought soap. I even take my soap with me when I travel because on the rare occasion when I've forgotten it, my skin protests by getting dry and itchy.

Processes

There are two methods for making soap—hot process and cold process. Every soap maker has their favorite method for a variety of reasons. I learned to make cold-process soap first and never saw the attraction of hot process, which takes more hands-on time. In the cold-process method, the oils and the lye are mixed together, and once the mixture reaches the thickness of a runny pudding, it is poured into molds. It continues to heat up and turns into soap in about twenty-four hours. The solid mass can then be sliced into bars, which are left to dry for about a month before being used.

In hot-process soap making, the ingredients are essentially "cooked," causing saponification to happen within a few hours. You need to stick around during this time and stir frequently so that the mixture does not bubble over. A slow cooker is often used in hot-process soap making. During the cooking process, the mixture goes through the gel phase, and quite a bit of the liquid will evaporate, which means the soap will dry faster. You can often tell the difference between soap made by hot process and by cold process when you see the finished bars of soap. Cold-process soap bars will have a smoother finish because the mixture was poured into the molds while still liquid, whereas hot-process bars will have a rougher surface because the mixture was pressed into molds when it was the consistency of mashed potatoes.

Being a busy person, I love the idea of making a batch of soap in little more than half an hour, pouring it into molds and then forgetting about it until the next day. But hot-process soap makers say their soap dries faster and that they can sell it or use it sooner. And although this is the case, the longer a soap dries, the longer it will last, so even hot-process bars can benefit from the month-long drying process that most cold-process soap makers follow.

If you are a fan of liquid soap, you will have to use a hot-process method to make it. The cold-process method can be used only for bar soap. Someone once suggested that I grate my bar soap and simply add water to create liquid soap. Unfortunately, when I did that, the soap developed a disgusting aroma within a couple of weeks. I didn't want to waste another bar of soap, so I didn't try again. And besides, I am quite happy with my bars of soap.

Safety

The need to use lye in making soap scares off some would-be soap makers. I understand completely because I felt the same way for a few months after deciding to learn to make soap. There are all sorts of horror stories on the Internet about people who have been injured by lye. Although we definitely need to be careful when making soap, we shouldn't be scared away from doing it.

The most important thing to remember is to make soap only when you will have an hour of uninterrupted time. If you have small children, don't do it when you are the only adult at home. If you have a cat that likes to jump on counters, lock it in another room. And don't make soap when you are rushed. Because that's when you are likely to make a mistake. Don't wear anything flowing or dangling, such as big sleeves or jewelry that could knock over a container or accidentally dip into the lye mixture or the unsaponified soap.

Some soap makers suit up with a lab coat, neoprene gloves, and goggles, and if you feel comfortable with this, it's not a bad idea. There are a couple of things to think about, though. If you are wearing a lab coat, jacket, or sweater, it should be a second layer. In other words, don't wear it over bare skin because an accidentally lye-soaked sleeve will be dragged across your skin when you remove the garment, thereby exposing more skin to the lye mixture. I personally don't wear gloves when making soap because they make me clumsy and I wind up dropping things, but if you can work with them, then use them, by all means. Do wear something to protect your eyes. Lye in your eye can blind you by causing chemical burns. I need reading glasses, and I have a pair of extra large ones that cover my eyes from my eyebrows to my cheeks that I use when making soap.

If you get a lye splatter on your skin, flush it with water, and if it still stings, rinse the skin with vinegar. If lye gets in your eye, you are supposed to flush with water for fifteen minutes and then go to the hospital emergency room. Once lye hits the eyeball, damage is done, so you need to rinse to minimize the damage and then have the eye checked by a medical professional.

Equipment

None of your soap-making equipment should be made of aluminum because it will react negatively with lye. Glass and stainless steel are the best bets. You can also use plastic and wood. However, they should be used exclusively for soap making because they are porous and will absorb the lye, fragrances, and soap. I accidentally used my plastic soap-making spatula for frosting a cake, and when I licked the spatula afterwards, it tasted like I had just licked a bar of soap!

- **Digital scale:** Yes, you really need a digital scale. Although people did make soap for thousands of years before digital scales were invented, the quality of the soap varied. The difference between a great batch of gentle soap and a caustic batch of soap can be made by as little as half an ounce of an ingredient in a small batch, and that isn't something you can eyeball in a measuring cup.
- **Pot:** A glass or stainless steel 3-quart pot will be needed for the recipes in this book. If you want to double them, you will need a 5-quart pot for melting the oils and mixing the final ingredients.
- **Pitcher:** A 2-quart pitcher or mixing bowl with handle is best for mixing the frozen goat milk and the lye together. The handle makes it easier to lift and pour the lye mixture.
- **Glass measuring cup:** You will need this to hold the lye and the essential oil for weighing. A glass measuring cup has a wide mouth that you are unlikely to miss when pouring lye into it, and there is a handle for safety when lifting and pouring.

Commercial molds create soap bars in a variety of shapes.

This soap mold was made with 1-inch by 3.5-inch wood. Although it can be put together with screws, we put ours together with extra-long bolts so that we can take it apart if necessary. The soap should slip out easily if the mold is lined with freezer paper, but if enough soap leaks out to make the loaf stick to the mold, it's nice to be able to take it apart easily.

- **Thermometer:** A standard cooking thermometer will work for making soap as long as it reads between 90°F and 140°F.
- **Utensils:** A spatula and a long-handled spoon are needed. These can be made from plastic or wood and should be used only for making soap.
- **Stick blender (immersion blender):** Unless you want to spend hours stirring, you need a stick blender. A mixer does not work as fast.
- **Freezer paper:** Lining the mold with freezer paper makes removing the soap easier. I initially used wax paper, and although it does work, it is not as durable as freezer paper, and wax paper is more challenging to remove from the soap because it tends to tear.
- **Molds:** You can buy soap molds, or you can use containers that you already have. Plastic storage containers, cardboard potato chip cans, or a heavy-duty shoe box lined with freezer paper will work.
- **Vinegar:** Lye is a base (or alkali) and vinegar is an acid, which neutralizes lye. In the event you spill dissolved lye or unsaponified soap on your skin, flush with water, and if the area still burns, rinse it with vinegar.

Ingredients

To make soap, you need only oil, lye, and water, but by using milk instead of water and by adding a variety of other ingredients, you can make soap that is moisturizing or astringent or even exfoliating.

- **Frozen milk:** Raw milk can be used for making soap, and it's the only milk I've used. But whether it is raw or pasteurized, the milk does need to be frozen. When lye hits liquid, it starts to react and heat up, and it will heat the milk quickly. Some instructions recommend placing the pitcher of milk in a sink of cold water to slow down the heating reaction when the lye is added, and if the milk you are using is slushy (or less frozen), it is a good idea. But if you are using chunks of frozen milk, you can add the lye and stir while the container is sitting on the counter. I freeze milk in freezer bags in the quantities I need for making a batch of soap (13 ounces or 26 ounces). I lay the bags flat for freezing so that I can take them out of the freezer within half an hour of needing them for making soap and break the milk into chunks. You can also freeze the milk in ice cube trays and weigh it before making the soap. Unlike the oils, which must be weighed precisely, the amount of milk you use is not critical. Using more milk will result in a bar of soap that takes a little longer to dry, whereas using a little less milk will result in a bar that takes less time to dry.

- **Lye:** You cannot make soap without lye! I cringe whenever I see an advertisement for homemade soap made without lye because it is impossible to make. You must have lye for saponification to occur. Typically, "no lye" claims mean the soap maker did not make the soap from scratch. They purchased a melt-n-pour base, which is soap that someone else made with lye, and then they melted the soap, added color or fragrance, and made bars. Sodium cocoate or sodium palmate listed as an ingredient on a soap label means coconut oil or palm oil was saponified with sodium hydroxide (lye). Potassium cocoate means potassium hydroxide (also lye) was used for saponification.

    Because sodium hydroxide is one of the ingredients in crystal meth, it is becoming increasingly difficult to find. Do not buy anything labeled as a drain opener unless the label clearly states that it is 100 percent sodium hydroxide. If you can't find lye at your local grocery or home improvement store, you can order it online.

- **Oils:** Any type of oil can be used to make soap, but oils cannot be used interchangeably. Each oil has its own saponification value, which means it may require more or less lye than another oil to saponify. Historically, soap was made with whatever oils were available locally. In North America, it was often lard from pigs or tallow from cows. In the Mediterranean, it was olive oil. In tropical areas, it was palm or coconut oil. Although you can make soap from a single oil, it will not be as good as one that includes different oils. Each oil contributes its unique properties to the soap. For example, lard and olive oil are good moisturizers and coconut oil is a good cleanser and creates suds. Blending coconut oil and olive oil in a soap will create a cleansing bar that is sudsy and moisturizing. Lard and olive oil don't create suds, and without suds, most people today don't feel like they are getting clean, so you'll probably want some coconut oil in your soap. Although coconut oil is a great cleanser, it can be too drying for some people, so having some olive oil in your soap makes it more moisturizing. Unlike other oils, palm oil separates when melted, which means you either need to buy homogenized palm oil or melt the entire container of palm oil and stir before measuring the amount needed in the recipe.
- **Essential oils:** We like soap to have some type of fragrance. To add scent to a natural soap you need to use essential oils. Fragrance oils are a proprietary blend of whatever the manufacturer wanted to mix together to create the fragrance, and the blend will include natural and artificial ingredients. Essential oils can be used singly, or you can blend them to make your own signature fragrances with all natural ingredients. My favorite blend is lavender, ylang ylang, and grapefruit. Peppermint and rosemary is a popular combination, as is spearmint and lemon.
- **Herbs:** Adding herbs to cold-processed soap does not work well because the heat of saponification makes most herbs turn black. They

are essentially cooked and totally devoid of any fragrance in the end. You can sprinkle herbs on top of the soap after pouring it into the mold to create a unique appearance.

- **Oatmeal:** The term "colloidal oatmeal" on a cosmetic label means oatmeal that has been ground up, which is easy to do with a coffee grinder. Ground oats can be mixed into the soap to create a soap that scrubs the skin gently. Whole rolled oats can be sprinkled on top of your soap after you put it into the mold, but you really don't want whole rolled oats mixed throughout your soap because it would be more abrasive.

- **Other botanicals:** All sorts of natural things such as wheat germ, poppy seeds, and ground-up egg shells can be used. If you have chickens that lay brown eggs, they will create a white dust when ground, so it won't change the color of your soap.

- **Clay:** Clays from around the world are available from soap-making supply companies. Each clay has unique properties and affect different skin types in different ways.

Step-by-Step Soap Making

Read through the list of equipment and ingredients and make sure you have assembled everything you will need before you get started. The process demands close attention and goes quickly once the lye is mixed in, so it is important to have everything within arm's reach and ready to go.

Cold-process soap making is a two-phase process where the oil phase and the water or milk phase are combined until they are emulsified, poured into molds, and eventually saponified, which means turned into soap.

1. If the recipe includes any oils that are solid at room temperature, such as palm oil or cocoa butter, weigh them and put them in a 3-quart stainless steel pot on the stove. Heat the solid oils on low just until they melt and then turn off the heat. Weigh the liquid oils and add them to the melted oils and stir. This usually brings the temperature into the desired range, which is 110°F to 120°F. If the temperature is higher than 120°F, wait until the temperature comes down before moving on to the next step.

2. Eye protection must be worn for this step, so put it on now. Place the frozen goat milk chunks into the 2-quart pitcher. Weigh the lye, and slowly add it to the frozen goat milk. Stir the mixture gently until the lye is completely dissolved. The lye will melt the frozen milk within a few minutes. You must ALWAYS add the lye to the liquid. Pouring liquid into lye can cause a violent eruption. When the milk has melted and the lye has dissolved the temperature of the mixture should be between 100°F and 120°F. The more frozen the milk is when you add the lye, the closer you will be to this temperature range. If the temperature goes above 120°F, wait until it is down to 120°F before moving on to the next step. Some sources say the oil mixture and lye mixture should be the same temperature, but that is not important. When learning to make soap, you want to keep the temperatures below 120°F because higher temperatures accelerate trace, which is stressful for an experienced soap maker to deal with and is a big headache for a novice!

If the goat milk chunks are solidly frozen, the lye and milk mixture will be a pretty pastel yellow. The less frozen the milk, the darker the mixture will be. In fact, if the milk is mostly thawed, the mixture will be an angry orange, and there will be little white chunks of milk fat floating in the mixture. It's not pretty, but it works. Don't panic. Make a mental note of how thawed the milk was, and don't do it again.

Oh, No! My Milk Melted!

---

3. Carefully and slowly pour the lye and milk mixture into the oil mixture. Using the stick blender, begin mixing the lye solution and the oils. If you are adding oatmeal, egg shells, clay, or essential oils, do it now. The warmer the temperature of the mixture, the faster the mixture will reach trace, which is when it is the consistency of a runny pudding. When you lift the stick blender and drops and dribbles sit on the surface rather than disappear into the liquid, it is time to pour it into the mold. If the mixture gets to the consistency of mashed potatoes, it has seized and will be too thick to pour. You will need

to spoon it up and mash it into a mold. It won't be pretty, but it will still be soap.

4. After pouring the mixture into a mold, cover the surface with freezer paper to keep ash from forming. Ash is harmless, but it worries some people, who incorrectly assume it's lye, which can be a challenge if you will be selling your soap.

5. Let the soap sit, covered only by the freezer paper—no additional insulation is needed—for at least 24 hours before unmolding and slicing. All of the recipes in this book are pretty forgiving, and you can slice the soap several days later if you don't have time to do it sooner. But at some point, most soaps will become too dry and hard to slice, so don't wait too long.

6. Place the bars on a wire rack or shelf to air dry for three or four weeks before using. Although you can use them sooner, they will last longer if given the proper amount of drying time.

❖ Butter Bar

This is a great soap for someone who has extremely dry skin. When I first developed this recipe, I started using it on my face, which did not have dry skin, and within a week my skin broke out as if I were a teenager. I still love it as a hand and body soap in the middle of our cold, dry winters. I usually make this soap unscented because many people with dry skin tend to have allergy issues, but you can add an essential oil for fragrance if you like.

2 ounces castor oil	2 ounces lanolin
4 ounces cocoa butter	13 ounces frozen goat milk
12 ounces coconut oil	6 ounces lye
20 ounces olive oil	2 ounces essential oil
4 ounces shea butter	

Makes approximately 11 five-ounce bars.

❖ Facial Soap for Oily Skin

I made this for my youngest daughter when she was a teenager and started to get pimples on her face and her back. It includes Australian green clay, which reputedly draws out oil from skin, and I added these particular essential oils because they tend to be astringent, especially the lemongrass.

4 ounces cocoa butter

16 ounces olive oil

8 ounces sunflower oil

6 ounces coconut oil

6 ounces palm oil

2 ounces castor oil

2 ounces grapeseed oil

13 ounces frozen goat milk

6 ounces lye

2 tablespoons Australian green clay

1 ounce lemongrass essential oil

½ ounce lime essential oil

½ ounce grapefruit essential oil

Makes approximately 11 five-ounce bars.

❖ Unscented Mocha Java

I first made this soap using non-deodorized cocoa butter in 2003 thinking that it would smell like coffee and chocolate. When it didn't, I decided never to make it again. But one of my customers fell in love with it and requests about ten bars every fall, so I keep making it, and other customers have also fallen in love with it. It's especially good for cleaning your hands after you have been dicing onions or handling a stinky buck. It is important to allow the coffee grounds to air dry before using them to make this soap, or the bars will be spongy.

24 ounces olive oil

8 ounces coconut oil

8 ounces palm oil

4 ounces cocoa butter

6 ounces lye

13 ounces milk

2–3 tablespoons
    dried coffee grounds

Makes approximately 11 five-ounce bars.

Goat milk soap is never very white, but it will be even darker if you add used coffee grounds, and you will also be able to see and feel the coffee grounds.

Final Thoughts

✦ ✦ ✦

The Last Thing I Learned From Coco

When I completed the first draft of this book and sent it to my publisher, I knew it was incomplete without a few final thoughts, but at that time, I wasn't sure what else I needed to say. While the manuscript was being edited, however, I realized what was missing.

After a two-hour drive in the middle of the night to the University of Illinois Veterinary Teaching Hospital, Coco died following the birth of quintuplets, and for only the second time in eleven years, I questioned why we were producing our own food rather than simply buying everything at the grocery store. Of course, it would be easier, but at that moment, I felt it would be so much less painful as well. In fact, many people ask me if it's hard when an animal dies, regardless of whether it was slaughtered for food or died of natural causes.

Death is probably the most difficult thing that anyone ever has to deal with when living this lifestyle, but it is inevitable. Someone once said that all livestock becomes dead stock someday. In modern society we don't see the circle of life on a regular basis, and it probably comes as no surprise that the first few deaths on a homestead can feel devastating. When it comes to goats and other livestock, the more you own, the more often you will have to deal with death. Our herd averages around eighteen milkers as well as a few older, retired does. It is unlikely that we will ever have a year when a goat does not die. Just as some humans are closer to us than others, there will be some goats that will have a special place in your heart. Coco was one of those goats for me. As I mentioned in an earlier story, she always thought she was my baby and would try to crawl into my lap whenever I sat down. Even as we sat in the back seat on the drive to the veterinary hospital when she was in labor, she was trying to edge her over-sized pregnant body into my lap.

In her nine years of life, Coco Chanel gave us twenty-seven kids and hundreds of gallons of milk. I think of her every day when I see her daughters Vera Wang and Nina Ricci. And I'm sure I'll think of her often as her newborn Bella Freud grows up and becomes a mother and a milk goat. I can point to aging blocks of cheddar and Gouda that include milk that Coco produced and that we'll be eating in the years to come.

Losing Coco was incredibly difficult, but when I consider the alternatives to producing our own meat and dairy products, I know this is the right path for us. We quit consuming factory-farmed meat in 1989 and were vegetarians for fourteen years before we moved out here and started producing our own meat. And as we have learned to make more and more of our own dairy products, we have been able to eliminate purchas-

ing factory-farmed milk and cheese. I sleep better at night knowing that the animals producing my food have names rather than ID numbers, and are loved and respected.

Raising goats and producing your own meat, milk, and other products is not the easiest lifestyle, and depending on how you manage your goats, it may or may not be the cheapest path. However, having your own goats can provide you with the freshest, most delicious dairy products as well as all-natural meat, fertilizer, soap, and even leather. You will also wind up with priceless memories. Even though it was painful to say good-bye, I wouldn't trade my nine years with Coco for anything.

Endnotes

1. M. Z. Ali Al Ahmad, Y. Chebloune, G. Chatagnon, J. L. Pellerin, and F. Fieni, "Is Caprine Arthritis Encephalitis Virus (CAEV) Transmitted Vertically to Early Embryo Development Stages (Morulae or Blastocyst) Via In Vitro Infected Frozen Semen?" Abstract, *Theriogenology* 77, no. 8 (May 2012), doi:10.1016/j.theriogenology.2011.12.012.

2. Centers for Disease Control and Prevention, "National Notifiable Diseases Surveillance System, 1993–2010," last updated November 12, 2012, cdc.gov/brucellosis/resources/surveillance.html.

3. "Docility," Igenity, accessed May 19, 2013, igenity.com/beef/profile /Docility.aspx.

4. Bridget Doyle, "City Extends Bidding Deadline for O'Hare Goat Herd," *Chicago Tribune*, September 26, 2012, articles.chicagotribune.com/2012 -09-26/news/ct-met-ohare-goats-deadline-extended-20120927_1_goat -animals-airport-property.

5. Dan Hoffman, "Mowing With Goats," *Google Official Blog*, May 01, 2009, googleblog.blogspot.com/2009/05/mowing-with-goats.html.

6. Sue Stehman and Mary Smith, "Goat Parasites: Management and Control" (revised September 2004), (presentation at ECA Symposium on Goat Health, June 3, 1995), accessed May 19, 2013, ansci.cornell.edu/goats /Resources/GoatArticles/GoatHealth/GoatParasites/Parasites-SM.pdf.

7. Steve Hart, "Dewormers and Dewormer Resistance," American Consortium for Small Ruminant Parasite Control, sheepandgoat.com/ACS RPC/Resources/dewormersHart.html.

8. Linda Coffey, Margo Hale, Tom Terrill, Jorge Mosjidis, Jim Miller, and Joan Burke, "Tools for Managing Internal Parasites in Small Ruminants: Copper Oxide Wire Particles," NCAT/ATTRA and Southern Consortium for Small Ruminant Parasite Control (2007), sheepandgoat.com /ACSRPC/Resources/PDF/COWP.pdf.

9. J. M. Burke, A. Wells, P. Casey, and R. M. Kaplan, "Herbal Dewormer

Fails to Control Gastrointestinal Nematodes in Goats," *Veterinary Parasitology* 160 (2009): 168–70, doi.org/10.1016/j.vetpar.2008.10.080.

10. Paula Simmons and Carol Ekarius, *Storey's Guide to Raising Sheep* (North Adams, MA: Storey Publishing, 2009).

11. Sue Stehman and Mary Smith, "Goat Parasites" (see note 6).

12. M. Z. Ali Al Ahmad et al., "Caprine Arthritis Encephalitis Virus" (see note 1).

13. John Matthews, *Diseases of the Goat*, 3rd ed. (West Sussex, UK: Wiley-Blackwell, 2009).

14. M. Soller and H. Angel, "Polledness and Abnormal Sex Ratios in Saanen Goats," *Journal of Heredity* (1964): 139–42.

Glossary

browse: Small trees, bushes, and leaves eaten by goats. A browser is an animal that prefers to eat those foods.

buck: A male goat.

buckling: A male kid.

cover: To breed, as in "The buck covered the doe."

cud: Food that is regurgitated from the goat's first stomach to be chewed again before going to the second stomach.

curd: The solid mass that is formed when making cheese; coagulated milk.

dehorn: To remove horns that have already started to grow.

disbud: To burn the horn buds so that horns don't grow.

dam: A mother goat.

doe: A female goat.

doeling: A female kid.

dry doe: An adult female goat that is not in milk.

dry-off: To stop milking a goat; the end of lactation.

flocculation: The point at which milk begins to turn into curds.

forage: Small trees and bushes.

freshen: To start making milk; used synonymously with "give birth."

horn buds: Bumps on the top of a kid's head that will grow into horns; not to be confused with polled bumps.

inflations: The part of a milking machine that goes over the doe's teats.

in-milk: A doe that is making milk.

kid: A baby goat.

poll: The top of a goat's head.

polled: A goat that does not have the genetic ability to grow horns but will grow small bumps on top of its head.

rumen: A goat's first stomach.

ruminant: An animal with four stomachs that chews its cud.

saponification: When a mixture of oils, lye, and milk turns into soap.

scurs: Small bits of horn that may grow after disbudding.

settle: To get pregnant.

sire: A father goat; "to sire" is to father kids, as in "The buck sired four doelings."

wether: A castrated male goat.

Suggested Reading

Books

Caldwell, Gianaclis and Ricki Carroll. *Mastering Artisan Cheesemaking: The Ultimate Guide for Home-Scale and Market Producers*. Chelsea Green Publishing, 2012.

Leverentz, James R. *The Complete Idiot's Guide to Cheese Making*. Alpha Books, 2010.

Matthews, John. *Diseases of the Goat*. 3rd ed. West Sussex, UK: Wiley-Blackwell, 2009.

Smith, Mary C. and David M. Sherman, *Goat Medicine*. 2nd ed. Wiley-Blackwell, 2009.

Solaiman, Sandra G. *Goat Science and Production*. Wiley-Blackwell, 2010.

Websites

American Consortium for Small Ruminant Parasite Control, wormcontrol.org

American Dairy Goat Association, adga.org

American Goat Society, americangoatsociety.com

Goat Management, Cornell University, Animal Science Department, ansci.cornell.edu/goats/resources_list.html

Maryland Small Ruminant Page, sheepandgoat.com

Recipe Index

Index

277

About the Author

DEBORAH NIEMANN is a homesteader, writer, and self-sufficiency expert. In 2002, she relocated her family from the suburbs of Chicago to a 32-acre parcel on a creek "in the middle of nowhere." Together, they built their own home and began growing the majority of their own food. Sheep, pigs, cattle, goats, chickens, and turkeys supply meat, eggs, and dairy products, while an organic garden and orchard provides fruit and vegetables. A highly sought-after speaker and workshop leader, Deborah presents extensively on topics including soap-making, bread-baking, cheese-making, composting and home-schooling. She is also the author of *Homegrown & Handmade* and *Ecothrifty*.

If you have enjoyed *Raising Goats Naturally*, you might also enjoy other

BOOKS TO BUILD A NEW SOCIETY

Our books provide positive solutions for people who want to
make a difference. We specialize in:

Sustainable Living • Green Building • Peak Oil
Renewable Energy • Environment & Economy
Natural Building & Appropriate Technology
Progressive Leadership • Resistance and Community
Educational & Parenting Resources

New Society Publishers

ENVIRONMENTAL BENEFITS STATEMENT

New Society Publishers has chosen to produce this book on recycled paper made
with **100% post consumer waste,** processed chlorine free, and old growth free.

For every 5,000 books printed, New Society saves the following resources:[1]

38	Trees
3,429	Pounds of Solid Waste
3,773	Gallons of Water
4,922	Kilowatt Hours of Electricity
6,234	Pounds of Greenhouse Gases
27	Pounds of HAPs, VOCs, and AOX Combined
9	Cubic Yards of Landfill Space

[1]Environmental benefits are calculated based on research done by the Environmental Defense Fund
and other members of the Paper Task Force who study the environmental impacts of the paper
industry.

For a full list of NSP's titles, please call 1-800-567-6772 *or check out our website* at:

www.newsociety.com